T0259703

Mechanics of Soft Materials

Konstantin Volokh

Mechanics of Soft Materials

 Springer

Konstantin Volokh
Faculty of Civil and Environmental
    Engineering
Technion - Israel Institute of Technology
Haifa
Israel

ISBN 978-981-10-9396-8   ISBN 978-981-10-1599-1   (eBook)
DOI 10.1007/978-981-10-1599-1

Printed on acid-free paper

This Springer imprint is published by Springer Nature
The registered company is Springer Science+Business Media Singapore Pte Ltd.

To Zhanna and Yurii Volokh

Preface

This book is an outgrowth of lecture notes of a graduate course on mechanics of soft materials that I have been teaching since 2009. The interest in the mechanics of soft materials is triggered by the development of new engineering and biomedical technologies. Mechanical behavior of soft materials is strongly nonlinear from both physical (constitutive equations) and geometrical (large deformations) standpoints and the standard texts on mechanics of materials are not enough in this case. The nonlinearities make the subject challenging, yet rich and exciting.

In writing this book I tried to help the reader in starting her own research journey as quickly and painlessly as possible. For this purpose, I elaborated excessively on some formulas and included practically important numerical examples. Isotropic and anisotropic hyperelasticity is the core of the mathematical modeling in mechanics of soft materials. However, thermo-, chemo-, electro-, and viscoelastic couplings are also often required and they are introduced in the book. Though this book is mostly about soft solids, the last chapter introduces a general Eulerian elasticity-fluidity framework, which is applicable to non-Newtonian fluids as well.

Finally, it is my great pleasure to acknowledge Ilia Volokh for the careful reading of the draft of the manuscript and thoughtful comments. During the writing process I also enjoyed the support from the Israel Science Foundation which came on time.

Haifa, Israel   Konstantin Volokh

Contents

Chapter 1
Tensors

In this chapter we introduce the concept of tensors which are the language of mechanics. Books are written on this topic and a newcomer might be interested in consulting the literature in advance. Nevertheless, all information about tensors necessary for understanding this book is collected bellow. We were not generous in choosing the information and we avoided any storage of results from tensor analysis which were not directly applicable to the problems considered in the present book.

A note concerning notation. Henceforth, we use lightface letters for scalars and boldface letters for vectors and tensors. Generally, yet not always, we use boldface lowercase letters for vectors and boldface uppercase letters for tensors.

1.1 Vectors

We assume that the reader is familiar with the basic vector algebra and analysis. The purpose of this section is to refresh memories and, more importantly, to introduce some new simple yet useful concepts of indicial summation (Einstein rule) and Kronecker and permutation (Levi-Civita) symbols. These concepts make manipulations with vectors and tensors easy.

1.1.1 Einstein Rule

Vectors are tensors of the first order (or rank), by definition, while scalars are zero-order tensors.

We consider Cartesian coordinate system with mutually orthogonal axes x_1, x_2, x_3 and unit *basis vectors* $\mathbf{e}_1$, $\mathbf{e}_2$, $\mathbf{e}_3$—Fig. 1.1. Within this coordinate system we define arbitrary vector $\mathbf{a}$ as follows

© Springer Science+Business Media Singapore 2016
K. Volokh, *Mechanics of Soft Materials*, DOI 10.1007/978-981-10-1599-1_1

Fig. 1.1 Vectors

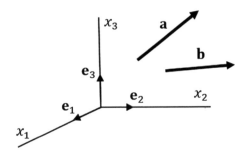

$$\mathbf{a} = a_1\mathbf{e}_1 + a_2\mathbf{e}_2 + a_3\mathbf{e}_3 = \sum_{i=1}^{3} a_i\mathbf{e}_i, \tag{1.1}$$

in which a_i is a vector coordinate or *component*.

Notation in (1.1) is excessive and it can be simplified by using the *Einstein rule*

$$\sum_{i=1}^{3} a_i\mathbf{e}_i = a_i\mathbf{e}_i, \tag{1.2}$$

which means that symbol of the sum $\sum_{i=1}^{3}$ can be dropped when the summation is performed over two *repeated* indices. Such indices are called *dummy* because they can be designated by any character

$$a_i\mathbf{e}_i = a_j\mathbf{e}_j = a_m\mathbf{e}_m. \tag{1.3}$$

1.1.2   Kronecker Delta

Using Einstein's rule we can write down the scalar or dot product of two vectors as follows

$$\mathbf{a} \cdot \mathbf{b} = (a_i\mathbf{e}_i) \cdot (b_j\mathbf{e}_j) = a_i b_j \mathbf{e}_i \cdot \mathbf{e}_j. \tag{1.4}$$

The scalar product of various couples of the basis vectors $\mathbf{e}_i \cdot \mathbf{e}_j$ can be designated via the *Kronecker delta* δ_{ij} for short notation

$$\mathbf{e}_i \cdot \mathbf{e}_j = \left\{ \begin{array}{l} 1, \ i = j \\ 0, \ i \neq j \end{array} \right\} \equiv \delta_{ij}. \tag{1.5}$$

Substituting (1.5) in (1.4) we obtain

$$\mathbf{a} \cdot \mathbf{b} = a_i b_j \mathbf{e}_i \cdot \mathbf{e}_j = a_i b_j \delta_{ij} = a_i b_i = a_1 b_1 + a_2 b_2 + a_3 b_3, \tag{1.6}$$

where

$$b_j \delta_{ij} = b_1 \delta_{i1} + b_2 \delta_{i2} + b_3 \delta_{i3} = b_i. \tag{1.7}$$

The latter identity can be interpreted as a manipulation allowing to "replace" index j by index i: $b_j \delta_{ij} = b_i$. Very useful.

By using the dot product of vector $\mathbf{a}$ and basis vector $\mathbf{e}_i$ we find the corresponding vector component

$$\mathbf{a} \cdot \mathbf{e}_i = (a_j \mathbf{e}_j) \cdot \mathbf{e}_i = a_j \delta_{ij} = a_i. \tag{1.8}$$

1.1.3 Permutation Symbol

The Kronecker delta was introduced via the scalar products of the Cartesian basis vectors. It is also very convenient to introduce the *permutation symbol* ε_{ijk} by using the triple product of various basis vectors

$$\mathbf{e}_i \cdot (\mathbf{e}_j \times \mathbf{e}_k) = \begin{cases} 1, & ijk = 123;\ 231;\ 312 \\ -1, & ijk = 321;\ 213;\ 132 \\ 0, & ijk = \cdots \end{cases} \equiv \varepsilon_{ijk}. \tag{1.9}$$

The permutation symbol allows us, for example, to write the components of the vector product in a short way

$$(\mathbf{a} \times \mathbf{b})_i = \mathbf{e}_i \cdot (\mathbf{a} \times \mathbf{b}) = \mathbf{e}_i \cdot \{(a_j \mathbf{e}_j) \times (b_k \mathbf{e}_k)\} = a_j b_k \mathbf{e}_i \cdot (\mathbf{e}_j \times \mathbf{e}_k) = \varepsilon_{ijk} a_j b_k. \tag{1.10}$$

It is important to note that there is no summation over index i in (1.10). Such index is called *free*. Computing (1.10) for varying i we get

$$c_1 = a_2 b_3 - a_3 b_2, \quad c_2 = a_3 b_1 - a_1 b_3, \quad c_3 = a_1 b_2 - a_2 b_1.$$

1.2 Second-Order Tensors

1.2.1 Definition

Vectors can be thought of as arrows in space. Unfortunately, there are no simple geometric objects to interpret second or higher-order tensors. There are two ways to think about tensors and define them. The first way is to define tensor as a *mapping*. For example, second-order tensors map vectors into vectors. Thus, tensor is an *action*. Alternatively, tensors can be defined as *objects* rather than actions. Tensors are objects created by the operation of *tensor product*.

For example, to define a second-order tensor we introduce the tensor product of basis vectors $\mathbf{e}_i \otimes \mathbf{e}_j$ which are called *basis dyads*. Their components can be calculated in Cartesian coordinates as follows

$$[\mathbf{e}_1 \otimes \mathbf{e}_1] = [\mathbf{e}_1][\mathbf{e}_1]^{\mathrm{T}} = \begin{bmatrix} 1 \\ 0 \\ 0 \end{bmatrix} [1\ 0\ 0] = \begin{bmatrix} 1\ 0\ 0 \\ 0\ 0\ 0 \\ 0\ 0\ 0 \end{bmatrix},$$

$$[\mathbf{e}_1 \otimes \mathbf{e}_2] = [\mathbf{e}_1][\mathbf{e}_2]^{\mathrm{T}} = \begin{bmatrix} 1 \\ 0 \\ 0 \end{bmatrix} [0\ 1\ 0] = \begin{bmatrix} 0\ 1\ 0 \\ 0\ 0\ 0 \\ 0\ 0\ 0 \end{bmatrix}, \qquad (1.11)$$

$$\vdots$$

$$[\mathbf{e}_3 \otimes \mathbf{e}_3] = [\mathbf{e}_3][\mathbf{e}_3]^{\mathrm{T}} = \begin{bmatrix} 0 \\ 0 \\ 1 \end{bmatrix} [0\ 0\ 1] = \begin{bmatrix} 0\ 0\ 0 \\ 0\ 0\ 0 \\ 0\ 0\ 1 \end{bmatrix}.$$

Here, we used brackets [...] to designate matrices of components.

Vectors are defined as linear combinations of basis vectors. By analogy with vectors, we define second-order tensors as linear combinations of the basis dyads

$$\begin{aligned} \mathbf{A} = {}& A_{11}\mathbf{e}_1 \otimes \mathbf{e}_1 + A_{12}\mathbf{e}_1 \otimes \mathbf{e}_2 + A_{13}\mathbf{e}_1 \otimes \mathbf{e}_3 \\ & + A_{21}\mathbf{e}_2 \otimes \mathbf{e}_1 + A_{22}\mathbf{e}_2 \otimes \mathbf{e}_2 + A_{23}\mathbf{e}_2 \otimes \mathbf{e}_3 \\ & + A_{31}\mathbf{e}_3 \otimes \mathbf{e}_1 + A_{32}\mathbf{e}_3 \otimes \mathbf{e}_2 + A_{33}\mathbf{e}_3 \otimes \mathbf{e}_3, \end{aligned} \qquad (1.12)$$

or

$$\mathbf{A} = \sum_{j=1}^{3} \sum_{i=1}^{3} A_{ij}\mathbf{e}_i \otimes \mathbf{e}_j, \qquad (1.13)$$

or, by using the Einstein rule,

$$\mathbf{A} = A_{ij}\mathbf{e}_i \otimes \mathbf{e}_j. \qquad (1.14)$$

In the matrix form we have

$$[\mathbf{A}] = A_{ij}[\mathbf{e}_i \otimes \mathbf{e}_j] = A_{ij}[\mathbf{e}_i][\mathbf{e}_j]^{\mathrm{T}}, \qquad (1.15)$$

or, after the substitution of the components of basis vectors,

$$\begin{aligned} [\mathbf{A}] = {}& A_{11} \begin{bmatrix} 1\ 0\ 0 \\ 0\ 0\ 0 \\ 0\ 0\ 0 \end{bmatrix} + A_{12} \begin{bmatrix} 0\ 1\ 0 \\ 0\ 0\ 0 \\ 0\ 0\ 0 \end{bmatrix} + A_{13} \begin{bmatrix} 0\ 0\ 1 \\ 0\ 0\ 0 \\ 0\ 0\ 0 \end{bmatrix} \\ & + A_{21} \begin{bmatrix} 0\ 0\ 0 \\ 1\ 0\ 0 \\ 0\ 0\ 0 \end{bmatrix} + A_{22} \begin{bmatrix} 0\ 0\ 0 \\ 0\ 1\ 0 \\ 0\ 0\ 0 \end{bmatrix} + A_{23} \begin{bmatrix} 0\ 0\ 0 \\ 0\ 0\ 1 \\ 0\ 0\ 0 \end{bmatrix} \end{aligned}$$

$$+ A_{31} \begin{bmatrix} 0\ 0\ 0 \\ 0\ 0\ 0 \\ 1\ 0\ 0 \end{bmatrix} + A_{32} \begin{bmatrix} 0\ 0\ 0 \\ 0\ 0\ 0 \\ 0\ 1\ 0 \end{bmatrix} + A_{33} \begin{bmatrix} 0\ 0\ 0 \\ 0\ 0\ 0 \\ 0\ 0\ 1 \end{bmatrix}$$

$$= \begin{bmatrix} A_{11}\ A_{12}\ A_{13} \\ A_{21}\ A_{22}\ A_{23} \\ A_{31}\ A_{32}\ A_{33} \end{bmatrix}. \qquad (1.16)$$

In the case of Cartesian coordinates the tensor can be interpreted as a matrix of its components. In the case of curvilinear coordinates the situation is subtler and various matrices of components can represent the same tensor. The latter case will be discussed below.

By analogy with the second-order tensors we can define higher-order tensors. For example, a *fourth-order tensor*[1] is formed by a multiple tensor product of basis vectors

$$\mathbb{C} = C_{mnij}\mathbf{e}_m \otimes \mathbf{e}_n \otimes \mathbf{e}_i \otimes \mathbf{e}_j. \qquad (1.17)$$

1.2.2 Multiplication of Tensors

A second-order tensor maps one vector into another as follows

$$\mathbf{c} = \mathbf{Ab} = (A_{ij}\mathbf{e}_i \otimes \mathbf{e}_j)(b_m\mathbf{e}_m) = A_{ij}b_m\mathbf{e}_i(\mathbf{e}_j \cdot \mathbf{e}_m) = A_{ij}b_m\mathbf{e}_i\delta_{jm} = A_{ij}b_j\mathbf{e}_i, \qquad (1.18)$$

or

$$c_i = A_{ij}b_j,$$

or

$$\begin{bmatrix} c_1 \\ c_2 \\ c_3 \end{bmatrix} = \begin{bmatrix} A_{11}\ A_{12}\ A_{13} \\ A_{21}\ A_{22}\ A_{23} \\ A_{31}\ A_{32}\ A_{33} \end{bmatrix} \begin{bmatrix} b_1 \\ b_2 \\ b_3 \end{bmatrix}.$$

It is worth noting that we use the following notational convention

$$(\mathbf{a} \otimes \mathbf{b})\mathbf{c} = (\mathbf{b} \cdot \mathbf{c})\mathbf{a}, \qquad (1.19)$$

while other authors use instead

$$(\mathbf{ab}) \cdot \mathbf{c} = (\mathbf{b} \cdot \mathbf{c})\mathbf{a}. \qquad (1.20)$$

In the latter case the symbol of tensor product $\otimes$ is omitted, which elevates the importance of the dot symbol "$\cdot$".

---

[1] We use the blackboard letter $\mathbb{C}$ for the symbolic notation of the fourth-order tensor.

Product of two second-order tensors is defined as follows

$$\begin{aligned}
\mathbf{AD} &= (A_{ij}\mathbf{e}_i \otimes \mathbf{e}_j)(D_{mn}\mathbf{e}_m \otimes \mathbf{e}_n) \\
&= A_{ij}D_{mn}\mathbf{e}_i \otimes \mathbf{e}_n(\mathbf{e}_j \cdot \mathbf{e}_m) \\
&= A_{ij}D_{mn}\delta_{jm}\mathbf{e}_i \otimes \mathbf{e}_n \\
&= A_{ij}D_{jn}\mathbf{e}_i \otimes \mathbf{e}_n.
\end{aligned} \tag{1.21}$$

The sum over the repeated index j in this formula is also called *contraction*. Some authors write $\mathbf{A} \cdot \mathbf{D}$ instead of $\mathbf{AD}$.

Double dot product of two tensors is a scalar

$$\begin{aligned}
\mathbf{A} : \mathbf{D} &= (A_{ij}\mathbf{e}_i \otimes \mathbf{e}_j) : (D_{mn}\mathbf{e}_m \otimes \mathbf{e}_n) \\
&= A_{ij}D_{mn}(\mathbf{e}_i \cdot \mathbf{e}_m)(\mathbf{e}_j \cdot \mathbf{e}_n) \\
&= A_{ij}D_{mn}\delta_{im}\delta_{jn} \\
&= A_{ij}D_{ij} \\
&= A_{11}D_{11} + A_{12}D_{12} + \cdots + A_{33}D_{33}.
\end{aligned} \tag{1.22}$$

Here, we have the double contraction.

Similarly, the double dot product of the fourth- and second- order tensors is defined as follows

$$\begin{aligned}
\mathbb{C} : \mathbf{B} &= (C_{mnij}\mathbf{e}_m \otimes \mathbf{e}_n \otimes \mathbf{e}_i \otimes \mathbf{e}_j) : (B_{kl}\mathbf{e}_k \otimes \mathbf{e}_l) \\
&= C_{mnij}B_{kl}\mathbf{e}_m \otimes \mathbf{e}_n(\mathbf{e}_i \cdot \mathbf{e}_k)(\mathbf{e}_j \cdot \mathbf{e}_l) \\
&= C_{mnij}B_{kl}\mathbf{e}_m \otimes \mathbf{e}_n\delta_{ik}\delta_{jl} \\
&= C_{mnij}B_{ij}\mathbf{e}_m \otimes \mathbf{e}_n.
\end{aligned} \tag{1.23}$$

By using the double dot product we can calculate components of a second-order tensor as follows

$$\mathbf{e}_i \otimes \mathbf{e}_j : \mathbf{A} = \mathbf{e}_i \otimes \mathbf{e}_j : (A_{mn}\mathbf{e}_m \otimes \mathbf{e}_n) = A_{mn}(\mathbf{e}_i \cdot \mathbf{e}_m)(\mathbf{e}_j \cdot \mathbf{e}_n) = A_{ij}. \tag{1.24}$$

This is a "projection" of tensor $\mathbf{A}$ on the "direction" of dyad $\mathbf{e}_i \otimes \mathbf{e}_j$.

If projection of tensor $\mathbf{A}$ on $\mathbf{v} \otimes \mathbf{v}$ is not negative for any nonzero vector $\mathbf{v} \neq \mathbf{0}$

$$\mathbf{A} : \mathbf{v} \otimes \mathbf{v} \geq 0, \tag{1.25}$$

then tensor $\mathbf{A}$ is called *positive semi-definite*.

In the case of the strong inequality

$$\mathbf{A} : \mathbf{v} \otimes \mathbf{v} > 0, \tag{1.26}$$

the tensor is called *positive definite*.

Analogously, negative definite and semi-negative definite tensors can be defined. Very special and useful is the *second-order identity tensor* defined as follows

$$\mathbf{1} = \delta_{ij}\mathbf{e}_i \otimes \mathbf{e}_j = \mathbf{e}_1 \otimes \mathbf{e}_1 + \mathbf{e}_2 \otimes \mathbf{e}_2 + \mathbf{e}_3 \otimes \mathbf{e}_3. \qquad (1.27)$$

It enjoys a remarkable property

$$\mathbf{A}\mathbf{1} = \mathbf{1}\mathbf{A} = \mathbf{A}. \qquad (1.28)$$

The double-dot product of tensor $\mathbf{A}$ with the identity tensor introduces *trace* of the tensor

$$\text{tr}\mathbf{A} = \mathbf{A} : \mathbf{1} = A_{ij}\delta_{ij} = A_{ii} = A_{11} + A_{22} + A_{33}. \qquad (1.29)$$

The *transposed* second-order tensor is

$$\mathbf{A}^{\mathsf{T}} = (A_{ij}\mathbf{e}_i \otimes \mathbf{e}_j)^{\mathsf{T}} = A_{ij}\mathbf{e}_j \otimes \mathbf{e}_i = A_{ji}\mathbf{e}_i \otimes \mathbf{e}_j. \qquad (1.30)$$

Transposition allows us to additively decompose any second-order tensor into *symmetric* and anti-symmetric (*skew*) parts

$$\begin{aligned}
\mathbf{A} &= \mathbf{A}_{\text{sym}} + \mathbf{A}_{\text{skew}}, \\
\mathbf{A}_{\text{sym}} &= \frac{1}{2}(\mathbf{A} + \mathbf{A}^{\mathsf{T}}) = \mathbf{A}_{\text{sym}}^{\mathsf{T}}, \\
\mathbf{A}_{\text{skew}} &= \frac{1}{2}(\mathbf{A} - \mathbf{A}^{\mathsf{T}}) = -\mathbf{A}_{\text{skew}}^{\mathsf{T}}.
\end{aligned} \qquad (1.31)$$

The *inverse* second order tensor, $\mathbf{A}^{-1}$, is defined through the identity

$$\mathbf{A}^{-1}\mathbf{A} = \mathbf{A}\mathbf{A}^{-1} = \mathbf{1}. \qquad (1.32)$$

The inversion is possible when the tensor is not *singular*

$$\det \mathbf{A} \neq 0.$$

1.2.3 Eigenvalues and Eigenvectors of Tensors

Tensors have their intrinsic properties which are represented by *eigenvalues* and *eigenvectors*. These properties are revealed by the solution of the *eigenproblem* for a second-order tensor $\mathbf{A}$. The *eigenvalue* (principal or characteristic or proper value) ζ and the *eigenvector* (principal or characteristic or proper vector) $\mathbf{n}$ of the tensor are defined by the following equation

$$\mathbf{A}\mathbf{n} = \zeta\mathbf{n}, \qquad (1.33)$$

or, in a matrix form,

$$\begin{bmatrix} A_{11} - \zeta & A_{12} & A_{13} \\ A_{21} & A_{22} - \zeta & A_{23} \\ A_{31} & A_{32} & A_{33} - \zeta \end{bmatrix} \begin{bmatrix} n_1 \\ n_2 \\ n_3 \end{bmatrix} = \begin{bmatrix} 0 \\ 0 \\ 0 \end{bmatrix}. \tag{1.34}$$

Thus, eigenvector $\mathbf{n}$ defines a characteristic direction that is not changing under the tensor mapping $\mathbf{An}$ and a unit length in this direction is scaled by a factor equal to eigenvalue ζ.

Equation (1.33) possesses a nontrivial solution when the determinant of the coefficient matrix, called *characteristic polynomial*, is singular

$$\det[\mathbf{A} - \zeta\mathbf{1}] = -\zeta^3 + \zeta^2 I_1(\mathbf{A}) - \zeta I_2(\mathbf{A}) + I_3(\mathbf{A}) = 0. \tag{1.35}$$

Here the *principal invariants* of tensor $\mathbf{A}$ are introduced

$$\begin{aligned} I_1(\mathbf{A}) &= A_{11} + A_{22} + A_{33} = \text{tr}\mathbf{A}, \\ I_2(\mathbf{A}) &= \frac{1}{2}\{(\text{tr}\mathbf{A})^2 - \text{tr}(\mathbf{A}^2)\}, \\ I_3(\mathbf{A}) &= \det\mathbf{A}. \end{aligned} \tag{1.36}$$

For symmetric tensor $\mathbf{A} = \mathbf{A}^T$ all roots of (1.35), $\zeta_1, \zeta_2, \zeta_3$, are real and it is possible to find three mutually *orthogonal* principal directions corresponding to the roots. The unit vectors in principal directions $\mathbf{n}^{(1)}, \mathbf{n}^{(2)}, \mathbf{n}^{(3)}$ obey the *orthonormality* conditions

$$\mathbf{n}^{(i)} \cdot \mathbf{n}^{(j)} = \delta_{ij}. \tag{1.37}$$

Projecting tensor $\mathbf{A}$ on the new coordinate axes coinciding with the principal directions we get

$$\mathbf{n}^{(i)} \otimes \mathbf{n}^{(j)} : \mathbf{A} = \mathbf{n}^{(i)} \cdot \mathbf{A}\mathbf{n}^{(j)} = \zeta_j \mathbf{n}^{(i)} \cdot \mathbf{n}^{(j)} = \zeta_j \delta_{ij}, \quad \text{(no sum)}. \tag{1.38}$$

Thus, only diagonal components of the tensor are nonzero and they are equal to the eigenvalues. Consequently, tensor $\mathbf{A}$ can enjoy the *spectral decomposition*

$$\mathbf{A} = \zeta_1 \mathbf{n}^{(1)} \otimes \mathbf{n}^{(1)} + \zeta_2 \mathbf{n}^{(2)} \otimes \mathbf{n}^{(2)} + \zeta_3 \mathbf{n}^{(3)} \otimes \mathbf{n}^{(3)}, \tag{1.39}$$

if $\zeta_1 \neq \zeta_2 \neq \zeta_3$, or

$$\mathbf{A} = (\zeta_1 - \zeta_2)\mathbf{n}^{(1)} \otimes \mathbf{n}^{(1)} + \zeta_2 \mathbf{1}, \tag{1.40}$$

if $\zeta_1 \neq \zeta_2 = \zeta_3$, or

$$\mathbf{A} = \zeta_1 \mathbf{1}, \tag{1.41}$$

if $\zeta_1 = \zeta_2 = \zeta_3$.

Based on the spectral decomposition it is convenient to introduce the logarithm and the square root of a symmetric positive definite (or positive semi-definite) tensor, $\zeta_i > 0$ (or $\zeta_i \geq 0$),

$$\ln\mathbf{A} = \sum_{k=1}^{3} (\ln\zeta_k)\mathbf{n}^{(k)} \otimes \mathbf{n}^{(k)}, \tag{1.42}$$

$$\sqrt{\mathbf{A}} = \sum_{k=1}^{3} \sqrt{\zeta_k}\,\mathbf{n}^{(k)} \otimes \mathbf{n}^{(k)}. \tag{1.43}$$

The spectral decomposition also allows us to calculate the principal invariants in the following form

$$\begin{aligned} I_1(\mathbf{A}) &= \zeta_1 + \zeta_2 + \zeta_3, \\ I_2(\mathbf{A}) &= \zeta_1\zeta_2 + \zeta_1\zeta_3 + \zeta_2\zeta_3, \\ I_3(\mathbf{A}) &= \zeta_1\zeta_2\zeta_3. \end{aligned} \tag{1.44}$$

Finally, we derive the useful *Cayley–Hamilton* formula pre-multiplying (1.35) with $\mathbf{n}^{(i)}$ and accounting for $\mathbf{A}^a\mathbf{n}^{(i)} = \zeta_i^a\mathbf{n}^{(i)}$

$$-\mathbf{A}^3 + I_1(\mathbf{A})\mathbf{A}^2 - I_2(\mathbf{A})\mathbf{A} + I_3(\mathbf{A})\mathbf{1} = \mathbf{0}. \tag{1.45}$$

1.3 Tensor Functions

Tensors can be arguments of functions: $f(\mathbf{A})$. Let us calculate the differential of scalar function f with respect to tensor argument $\mathbf{A}$

$$df = \frac{\partial f}{\partial A_{ij}}dA_{ij}. \tag{1.46}$$

Here the components of the tensor increment can be written following the projection rule (1.24) as

$$dA_{ij} = \mathbf{e}_i \otimes \mathbf{e}_j : d\mathbf{A}, \tag{1.47}$$

and, substituting it in (1.46), we get

$$df = \frac{\partial f}{\partial A_{ij}}\mathbf{e}_i \otimes \mathbf{e}_j : d\mathbf{A}. \tag{1.48}$$

The latter equation prompts the definition of a symbolic derivative with respect to a second-order tensor

$$\frac{\partial f}{\partial \mathbf{A}} = \frac{\partial f}{\partial A_{ij}}\mathbf{e}_i \otimes \mathbf{e}_j. \tag{1.49}$$

Analogously, it is possible to define the derivative of a second-order tensor with respect to another second-order tensor

$$\frac{\partial \mathbf{A}}{\partial \mathbf{B}} = \frac{\partial \mathbf{A}}{\partial B_{ij}} \otimes \mathbf{e}_i \otimes \mathbf{e}_j = \frac{\partial A_{mn}}{\partial B_{ij}} \mathbf{e}_m \otimes \mathbf{e}_n \otimes \mathbf{e}_i \otimes \mathbf{e}_j. \tag{1.50}$$

For example, let us differentiate a second-order tensor with respect to itself

$$\frac{\partial \mathbf{A}}{\partial \mathbf{A}} = \frac{\partial A_{mn}}{\partial A_{ij}} \mathbf{e}_m \otimes \mathbf{e}_n \otimes \mathbf{e}_i \otimes \mathbf{e}_j = \delta_{mi}\delta_{nj}\mathbf{e}_m \otimes \mathbf{e}_n \otimes \mathbf{e}_i \otimes \mathbf{e}_j. \tag{1.51}$$

In the case of symmetric tensor, the symmetry should be preserved in the derivative

$$\frac{\partial \mathbf{A}}{\partial \mathbf{A}} = \frac{1}{2}\frac{\partial (\mathbf{A} + \mathbf{A}^{\mathrm{T}})}{\partial \mathbf{A}} = \frac{1}{2}(\delta_{mi}\delta_{nj} + \delta_{ni}\delta_{mj})\mathbf{e}_m \otimes \mathbf{e}_n \otimes \mathbf{e}_i \otimes \mathbf{e}_j. \tag{1.52}$$

Further important formulas are obtained by differentiation of the principal invariants

$$\begin{aligned}
\frac{\partial I_1(\mathbf{A})}{\partial \mathbf{A}} &= \frac{\partial A_{kk}}{\partial A_{ij}}\mathbf{e}_i \otimes \mathbf{e}_j = \delta_{ki}\delta_{kj}\mathbf{e}_i \otimes \mathbf{e}_j = \delta_{ij}\mathbf{e}_i \otimes \mathbf{e}_j = \mathbf{1}, \\
\frac{\partial I_2(\mathbf{A})}{\partial \mathbf{A}} &= \frac{1}{2}\frac{\partial (A_{kk}A_{ll} - A_{mn}A_{nm})}{\partial A_{ij}}\mathbf{e}_i \otimes \mathbf{e}_j = I_1(\mathbf{A})\mathbf{1} - \mathbf{A}_{\mathrm{sym}}.
\end{aligned} \tag{1.53}$$

The derivative of the third invariant is less trivial to find. The plan is to calculate it indirectly by using the following truncated Taylor series expansion

$$I_3(\mathbf{A} + d\mathbf{A}) = I_3(\mathbf{A}) + \frac{\partial I_3(\mathbf{A})}{\partial \mathbf{A}} : d\mathbf{A}. \tag{1.54}$$

Indeed, by a direct calculation with the help of (1.35) and neglecting higher order terms we have

$$\begin{aligned}
I_3(\mathbf{A} + d\mathbf{A}) &= \det(\mathbf{A} + d\mathbf{A}) \\
&= \det\mathbf{A}\det(\mathbf{A}^{-1}d\mathbf{A} - (-1)\mathbf{1}) \\
&= \det\mathbf{A}\{1 + I_1(\mathbf{A}^{-1}d\mathbf{A}) + I_2(\mathbf{A}^{-1}d\mathbf{A}) + I_3(\mathbf{A}^{-1}d\mathbf{A})\} \\
&= \det\mathbf{A} + (\det\mathbf{A})I_1(\mathbf{A}^{-1}d\mathbf{A}) \\
&= I_3(\mathbf{A}) + I_3(\mathbf{A})\mathbf{A}^{-\mathrm{T}} : d\mathbf{A}.
\end{aligned} \tag{1.55}$$

Comparing two previous equations we get finally

$$\frac{\partial I_3(\mathbf{A})}{\partial \mathbf{A}} = I_3(\mathbf{A})\mathbf{A}^{-\mathrm{T}}. \tag{1.56}$$

1.4 Tensor Fields

We turn to analysis of *tensor fields*, i.e. tensors determined and varying in the physical three-dimensional space.

1.4.1 Differential Operators

We *define* the following differential operators for scalars, vectors, and second-order tensors in Cartesian coordinates

$$
\begin{aligned}
\operatorname{grad}\varphi &= \frac{\partial\varphi}{\partial\mathbf{x}} = \frac{\partial\varphi}{\partial x_i}\mathbf{e}_i, \\[6pt]
\operatorname{grad}\mathbf{a} &= \frac{\partial\mathbf{a}}{\partial\mathbf{x}} = \frac{\partial\mathbf{a}}{\partial x_i}\otimes\mathbf{e}_i = \frac{\partial a_j}{\partial x_i}\mathbf{e}_j\otimes\mathbf{e}_i, \\[6pt]
\operatorname{div}\mathbf{a} &= \frac{\partial\mathbf{a}}{\partial x_i}\cdot\mathbf{e}_i = \frac{\partial a_j}{\partial x_i}\mathbf{e}_j\cdot\mathbf{e}_i = \frac{\partial a_i}{\partial x_i}, \\[6pt]
\operatorname{curl}\mathbf{a} &= \mathbf{e}_i\times\frac{\partial\mathbf{a}}{\partial x_i} = \mathbf{e}_i\times\mathbf{e}_j\frac{\partial a_j}{\partial x_i} = \varepsilon_{ijk}\frac{\partial a_j}{\partial x_i}\mathbf{e}_k, \\[6pt]
\operatorname{div}\mathbf{A} &= \frac{\partial\mathbf{A}}{\partial x_i}\mathbf{e}_i = \frac{\partial A_{mn}}{\partial x_i}(\mathbf{e}_m\otimes\mathbf{e}_n)\mathbf{e}_i = \frac{\partial A_{mn}}{\partial x_n}\mathbf{e}_m.
\end{aligned}
\tag{1.57}
$$

It is possible to use an alternative notation via the *nabla* vector-operator

$$
\nabla = \mathbf{e}_1\frac{\partial}{\partial x_1} + \mathbf{e}_2\frac{\partial}{\partial x_2} + \mathbf{e}_3\frac{\partial}{\partial x_3} = \mathbf{e}_i\frac{\partial}{\partial x_i}.
\tag{1.58}
$$

By using nabla we can rewrite the differential operators as follows

$$
\begin{aligned}
\operatorname{grad}\varphi &= \nabla\varphi = \frac{\partial\varphi}{\partial x_i}\mathbf{e}_i, \\[6pt]
\operatorname{grad}\mathbf{a} &= (\nabla\otimes\mathbf{a})^{\mathrm{T}} = \left(\mathbf{e}_i\otimes\frac{\partial\mathbf{a}}{\partial x_i}\right)^{\mathrm{T}} = \frac{\partial a_j}{\partial x_i}\mathbf{e}_j\otimes\mathbf{e}_i, \\[6pt]
\operatorname{div}\mathbf{a} &= \nabla\cdot\mathbf{a} = \mathbf{e}_i\cdot\frac{\partial\mathbf{a}}{\partial x_i} = \frac{\partial a_j}{\partial x_i}\mathbf{e}_i\cdot\mathbf{e}_j = \frac{\partial a_i}{\partial x_i}, \\[6pt]
\operatorname{curl}\mathbf{a} &= \nabla\times\mathbf{a} = \mathbf{e}_i\times\frac{\partial\mathbf{a}}{\partial x_i} = \mathbf{e}_i\times\mathbf{e}_j\frac{\partial a_j}{\partial x_i} = \varepsilon_{ijk}\frac{\partial a_j}{\partial x_i}\mathbf{e}_k, \\[6pt]
\operatorname{div}\mathbf{A} &= \nabla\cdot\mathbf{A}^{\mathrm{T}} = \mathbf{e}_i\cdot\frac{\partial\mathbf{A}^{\mathrm{T}}}{\partial x_i} = \frac{\partial A_{mn}}{\partial x_i}(\mathbf{e}_m\otimes\mathbf{e}_n)\mathbf{e}_i = \frac{\partial A_{mn}}{\partial x_n}\mathbf{e}_m.
\end{aligned}
\tag{1.59}
$$

The reader should be aware of the fact that the definitions of the differential operators are not standardized and they may vary from one author to another.

1.4.2 Integral Formulas

Most spatial integration formulas are based on the divergence theorem (Gauss–Green–Ostrogradskii), which is an important tool for transformation of volume, area, and line integrals. Its simplest version in one-dimensional case is the famous Newton–Leibniz rule—Fig. 1.2

$$\int_a^b \frac{dy}{dx}dx = (+1)y(b) + (-1)y(a) = n(b)y(b) + n(a)y(a).$$

In the three-dimensional case—Fig. 1.3—we can write (without proof)

$$\int \frac{\partial y}{\partial x_i}dV = \oint n_i dA, \tag{1.60}$$

where $y(\mathbf{x})$ is a spatial scalar field defined in volume V with the bounding surface A and outward unit normal $\mathbf{n}$.

A powerful consequence of (1.60) is the following formula

$$\int \frac{\partial B_{ij}}{\partial x_j}dV = \int \frac{\partial B_{i1}}{\partial x_1}dV + \int \frac{\partial B_{i2}}{\partial x_2}dV + \int \frac{\partial B_{i3}}{\partial x_3}dV$$
$$= \oint B_{i1}n_1 dA + \oint B_{i2}n_2 dA + \oint B_{i3}n_3 dA$$
$$= \oint B_{ij}n_j dA, \tag{1.61}$$

or, in a compact form,

$$\int \operatorname{div}\mathbf{B}dV = \oint \mathbf{B}\mathbf{n}dA. \tag{1.62}$$

Fig. 1.2 One-dimensional integration

Fig. 1.3 Three-dimensional integration

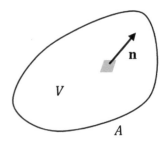

Fig. 1.4 Stokes integral
formula

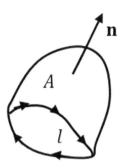

Similarly, one can obtain for scalar and vector

$$\int \text{grad}\varphi \, dV = \oint \varphi \mathbf{n} \, dA,$$

$$\int \text{div} \mathbf{a} \, dV = \oint \mathbf{a} \cdot \mathbf{n} \, dA. \tag{1.63}$$

Another useful formula is due to Stokes who connected the contour integral over curve l to surface A built on it—Fig. 1.4

$$\oint \mathbf{a} \cdot d\mathbf{x} = \int (\text{curl} \mathbf{a}) \cdot \mathbf{n} \, dA, \tag{1.64}$$

where $d\mathbf{x}$ is the infinitesimal element of the curve l.

It is remarkable that all mentioned formulas provide the relationships between integrals in different dimensions: volume–surface; surface–curve; curve–points.

1.5 Curvilinear Coordinates

Some problems are easier to solve in *curvilinear* rather than Cartesian coordinates. We consider curvilinear coordinates $\alpha^1, \alpha^2, \alpha^3$, which can be defined through the Cartesian coordinates x_1, x_2, x_3 and vice versa.

For example, in the case of *cylindrical* coordinates—Fig. 1.5—we have

$$\alpha^1 = r, \quad \alpha^2 = \varphi, \quad \alpha^3 = z, \tag{1.65}$$

and

$$x_1 = r \cos \varphi, \quad x_2 = r \sin \varphi, \quad x_3 = z, \tag{1.66}$$

Fig. 1.5 Cylindrical
coordinates

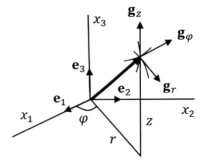

or, after inversion,

$$r = \sqrt{x_1^2 + x_2^2}, \quad \varphi = \arctan\frac{x_2}{x_1}, \quad z = x_3. \tag{1.67}$$

We define the *natural (co-variant)* basis vectors in curvilinear coordinates

$$\mathbf{s}_i = \frac{\partial \mathbf{x}}{\partial \alpha^i}, \tag{1.68}$$

which take following forms in the Cartesian basis

$$\mathbf{s}_r = \frac{\partial x_1}{\partial r}\mathbf{e}_1 + \frac{\partial x_2}{\partial r}\mathbf{e}_2 + \frac{\partial x_3}{\partial r}\mathbf{e}_3 = \cos\varphi\,\mathbf{e}_1 + \sin\varphi\,\mathbf{e}_2,$$

$$\mathbf{s}_\varphi = \frac{\partial x_1}{\partial \varphi}\mathbf{e}_1 + \frac{\partial x_2}{\partial \varphi}\mathbf{e}_2 + \frac{\partial x_3}{\partial \varphi}\mathbf{e}_3 = -r\sin\varphi\,\mathbf{e}_1 + r\cos\varphi\,\mathbf{e}_2, \tag{1.69}$$

$$\mathbf{s}_z = \frac{\partial x_1}{\partial z}\mathbf{e}_1 + \frac{\partial x_2}{\partial z}\mathbf{e}_2 + \frac{\partial x_3}{\partial z}\mathbf{e}_3 = \mathbf{e}_3.$$

We also define the *dual (contra-variant)* basis vectors

$$\mathbf{s}^i = \frac{\partial \alpha^i}{\partial \mathbf{x}}, \tag{1.70}$$

which take the following form in the Cartesian basis

$$\mathbf{s}^r = \frac{\partial r}{\partial x_1}\mathbf{e}_1 + \frac{\partial r}{\partial x_2}\mathbf{e}_2 + \frac{\partial r}{\partial x_3}\mathbf{e}_3 = \cos\varphi\,\mathbf{e}_1 + \sin\varphi\,\mathbf{e}_2,$$

$$\mathbf{s}^\varphi = \frac{\partial \varphi}{\partial x_1}\mathbf{e}_1 + \frac{\partial \varphi}{\partial x_2}\mathbf{e}_2 + \frac{\partial \varphi}{\partial x_3}\mathbf{e}_3 = -\frac{\sin\varphi}{r}\mathbf{e}_1 + \frac{\cos\varphi}{r}\mathbf{e}_2, \tag{1.71}$$

$$\mathbf{s}^z = \frac{\partial z}{\partial x_1}\mathbf{e}_1 + \frac{\partial z}{\partial x_2}\mathbf{e}_2 + \frac{\partial z}{\partial x_3}\mathbf{e}_3 = \mathbf{e}_3.$$

The natural and dual basis vectors are mutually orthogonal

$$\mathbf{s}^j \cdot \mathbf{s}_i = \frac{\partial \alpha^j}{\partial \mathbf{x}} \cdot \frac{\partial \mathbf{x}}{\partial \alpha^i} = \frac{\partial \alpha^j}{\partial \alpha^i} = \begin{Bmatrix} 1, & i = j \\ 0, & i \neq j \end{Bmatrix}. \tag{1.72}$$

Now, vectors have two representations in curvilinear coordinates

$$\mathbf{a} = a^i \mathbf{s}_i = a_i \mathbf{s}^i, \tag{1.73}$$

where

$$\begin{aligned} a^i &= \mathbf{a} \cdot \mathbf{s}^i, \\ a_i &= \mathbf{a} \cdot \mathbf{s}_i \end{aligned} \tag{1.74}$$

are *contra-* and *co-variant* components of the vector respectively.

Second-order tensors have four representations in curvilinear coordinates

$$\mathbf{A} = A^{ij}\mathbf{s}_i \otimes \mathbf{s}_j = A_{ij}\mathbf{s}^i \otimes \mathbf{s}^j = A^i_{\cdot j}\mathbf{s}_i \otimes \mathbf{s}^j = A_i^{\cdot j}\mathbf{s}^i \otimes \mathbf{s}_j, \tag{1.75}$$

where

$$\begin{aligned} A^{ij} &= \mathbf{A} : \mathbf{s}^i \otimes \mathbf{s}^j, \\ A_{ij} &= \mathbf{A} : \mathbf{s}_i \otimes \mathbf{s}_j, \\ A^i_{\cdot j} &= \mathbf{A} : \mathbf{s}^i \otimes \mathbf{s}_j, \\ A_i^{\cdot j} &= \mathbf{A} : \mathbf{s}_i \otimes \mathbf{s}^j \end{aligned} \tag{1.76}$$

are *contra-*, *co-*, *contra-co-*, and *co-contra-variant* components of the second-order tensor respectively.

The reader should note that upper indices (superscripts) correspond to the contra-variant quantities while the lower indices (subscripts) correspond to the co-variant ones. Remarkably, the Einstein summation rule is now applied to the repeated indices at different heights: lower-upper or co-contra.

If the basis vectors are mutually orthogonal then it is possible to normalize them as, for example, in the case of the cylindrical coordinates

$$\begin{aligned} \mathbf{g}_r &= \frac{\mathbf{s}_r}{|\mathbf{s}_r|} = \frac{\mathbf{s}^r}{|\mathbf{s}^r|} = \cos\varphi \mathbf{e}_1 + \sin\varphi \mathbf{e}_2, \\ \mathbf{g}_\varphi &= \frac{\mathbf{s}_\varphi}{|\mathbf{s}_\varphi|} = \frac{\mathbf{s}^\varphi}{|\mathbf{s}^\varphi|} = -\sin\varphi \mathbf{e}_1 + \cos\varphi \mathbf{e}_2, \\ \mathbf{g}_z &= \frac{\mathbf{s}_z}{|\mathbf{s}_z|} = \frac{\mathbf{s}^z}{|\mathbf{s}^z|} = \mathbf{e}_3. \end{aligned} \tag{1.77}$$

The normalized basis vectors allow introducing the so-called *physical components* of vectors and tensors with the same units

$$\mathbf{a} = a_r \mathbf{g}_r + a_\varphi \mathbf{g}_\varphi + a_z \mathbf{g}_z, \tag{1.78}$$

and

$$
\begin{aligned}
\mathbf{A} = {} & A_{rr} \mathbf{g}_r \otimes \mathbf{g}_r + A_{r\varphi} \mathbf{g}_r \otimes \mathbf{g}_\varphi + A_{rz} \mathbf{g}_r \otimes \mathbf{g}_z \\
& + A_{\varphi r} \mathbf{g}_\varphi \otimes \mathbf{g}_r + A_{\varphi\varphi} \mathbf{g}_\varphi \otimes \mathbf{g}_\varphi + A_{\varphi z} \mathbf{g}_\varphi \otimes \mathbf{g}_z \\
& + A_{zr} \mathbf{g}_z \otimes \mathbf{g}_r + A_{z\varphi} \mathbf{g}_z \otimes \mathbf{g}_\varphi + A_{zz} \mathbf{g}_z \otimes \mathbf{g}_z.
\end{aligned} \tag{1.79}
$$

By using the chain rule it is possible to calculate the differential operators in curvilinear coordinates

$$\text{grad}\,\mathbf{a} = \frac{\partial \mathbf{a}}{\partial x_i} \otimes \mathbf{e}_i = \frac{\partial \mathbf{a}}{\partial \alpha^j} \otimes \frac{\partial \alpha^j}{\partial x_i} \mathbf{e}_i = \frac{\partial \mathbf{a}}{\partial \alpha^j} \otimes \mathbf{s}^j,$$

$$\text{curl}\,\mathbf{a} = \mathbf{e}_i \times \frac{\partial \mathbf{a}}{\partial x_i} = \frac{\partial \alpha^j}{\partial x_i} \mathbf{e}_i \times \frac{\partial \mathbf{a}}{\partial \alpha^j} = \mathbf{s}^j \times \frac{\partial \mathbf{a}}{\partial \alpha^j}, \tag{1.80}$$

$$\text{div}\,\mathbf{A} = \frac{\partial \mathbf{A}}{\partial x_i} \mathbf{e}_i = \frac{\partial \mathbf{A}}{\partial \alpha^j} \frac{\partial \alpha^j}{\partial x_i} \mathbf{e}_i = \frac{\partial \mathbf{A}}{\partial \alpha^j} \mathbf{s}^j.$$

In the case of cylindrical coordinates we have, for example,

$$
\begin{aligned}
\text{grad}\,\mathbf{a} = {} & \frac{\partial \mathbf{a}}{\partial r} \otimes \mathbf{s}^r + \frac{\partial \mathbf{a}}{\partial \varphi} \otimes \mathbf{s}^\varphi + \frac{\partial \mathbf{a}}{\partial z} \otimes \mathbf{s}^z \\
= {} & \frac{\partial \mathbf{a}}{\partial r} \otimes \mathbf{g}_r + \frac{1}{r} \frac{\partial \mathbf{a}}{\partial \varphi} \otimes \mathbf{g}_\varphi + \frac{\partial \mathbf{a}}{\partial z} \otimes \mathbf{g}_z.
\end{aligned} \tag{1.81}
$$

It should not be overlooked in calculating the derivatives of vectors and tensors that the natural and dual and physical *basis vectors depend on coordinates*. In the considered case of cylindrical coordinates we have the following derivatives of the physical basis vectors

$$\frac{\partial \mathbf{g}_r}{\partial r} = \frac{\partial \mathbf{g}_\varphi}{\partial r} = \frac{\partial \mathbf{g}_z}{\partial r} = \frac{\partial \mathbf{g}_z}{\partial \varphi} = \frac{\partial \mathbf{g}_r}{\partial z} = \frac{\partial \mathbf{g}_\varphi}{\partial z} = \frac{\partial \mathbf{g}_z}{\partial z} = \mathbf{0},$$

$$\frac{\partial \mathbf{g}_r}{\partial \varphi} = \mathbf{g}_\varphi, \quad \frac{\partial \mathbf{g}_\varphi}{\partial \varphi} = -\mathbf{g}_r. \tag{1.82}$$

In addition to the considered cylindrical coordinates it is useful to list the basic relationships for *spherical coordinates*—Fig. 1.6.

Fig. 1.6 Spherical
coordinates

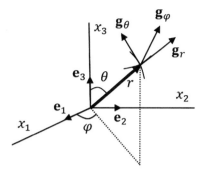

In this case we have

$$\alpha^1 = r, \quad \alpha^2 = \theta, \quad \alpha^3 = \varphi, \tag{1.83}$$

and

$$x_1 = r \cos \varphi \sin \theta, \quad x_2 = r \sin \varphi \sin \theta, \quad x_3 = r \cos \varphi, \tag{1.84}$$

or, after inversion,

$$r = \sqrt{x_1^2 + x_2^2 + x_3^2}, \quad \theta = \frac{x_3}{\sqrt{x_1^2 + x_2^2 + x_3^2}}, \quad \varphi = \arctan \frac{x_2}{x_1}. \tag{1.85}$$

This coordinate system is also orthogonal as in the case of cylindrical coordinates and the orthonormal physical basis vectors can be introduced

$$\begin{aligned}
\mathbf{g}_r &= \cos \varphi \sin \theta \mathbf{e}_1 + \sin \varphi \sin \theta \mathbf{e}_2 + \cos \theta \mathbf{e}_3, \\
\mathbf{g}_\theta &= \cos \varphi \cos \theta \mathbf{e}_1 + \sin \varphi \cos \theta \mathbf{e}_2 - \sin \theta \mathbf{e}_3, \\
\mathbf{g}_\varphi &= -\sin \varphi \mathbf{e}_1 + \cos \varphi \mathbf{e}_2.
\end{aligned} \tag{1.86}$$

The gradient operator is calculated as follows

$$\mathrm{grad}\, \mathbf{a} = \frac{\partial \mathbf{a}}{\partial r} \otimes \mathbf{g}_r + \frac{1}{r} \frac{\partial \mathbf{a}}{\partial \theta} \otimes \mathbf{g}_\theta + \frac{1}{r \sin \theta} \frac{\partial \mathbf{a}}{\partial \varphi} \otimes \mathbf{g}_\varphi, \tag{1.87}$$

and the derivatives of the basis vectors are

$$\begin{aligned}
&\frac{\partial \mathbf{g}_r}{\partial r} = \frac{\partial \mathbf{g}_\theta}{\partial r} = \frac{\partial \mathbf{g}_\varphi}{\partial r} = \frac{\partial \mathbf{g}_\varphi}{\partial \theta} = \mathbf{0}, \quad \frac{\partial \mathbf{g}_r}{\partial \theta} = \mathbf{g}_\theta, \quad \frac{\partial \mathbf{g}_\theta}{\partial \theta} = -\mathbf{g}_r, \\
&\frac{\partial \mathbf{g}_r}{\partial \varphi} = \mathbf{g}_\varphi \sin \theta, \quad \frac{\partial \mathbf{g}_\theta}{\partial \varphi} = \mathbf{g}_\varphi \cos \theta, \quad \frac{\partial \mathbf{g}_\varphi}{\partial \varphi} = -\mathbf{g}_r \sin \theta - \mathbf{g}_\theta \cos \theta.
\end{aligned} \tag{1.88}$$

1.6   Exercises

1. Prove:

$$\varepsilon_{skt}\varepsilon_{mnp} = \det \begin{bmatrix} \delta_{sm} & \delta_{sn} & \delta_{sp} \\ \delta_{km} & \delta_{kn} & \delta_{kp} \\ \delta_{tm} & \delta_{tn} & \delta_{tp} \end{bmatrix},$$

$$\varepsilon_{skt}\varepsilon_{snp} = \delta_{kn}\delta_{tp} - \delta_{kp}\delta_{tn},$$

$$\varepsilon_{skt}\varepsilon_{skp} = 2\delta_{tp},$$

$$\varepsilon_{skt}\varepsilon_{skt} = 6.$$

(1.89)

2. Prove (1.28).
3. Prove for second-order tensors $\mathbf{A}$, $\mathbf{B}$:

$$\det \mathbf{A} = \frac{1}{6}\varepsilon_{stp}\varepsilon_{ijk}A_{si}A_{tj}A_{pk},$$

$$\det(\mathbf{A}^{-1}) = (\det \mathbf{A})^{-1},$$

$$(\mathbf{AB})^{-1} = \mathbf{B}^{-1}\mathbf{A}^{-1},$$

$$(\mathbf{A}^{\mathrm{T}})^{-1} = (\mathbf{A}^{-1})^{\mathrm{T}} \equiv \mathbf{A}^{-\mathrm{T}},$$

$$\mathrm{tr}(\mathbf{AB}) = \mathbf{A}^{\mathrm{T}} : \mathbf{B}.$$

(1.90)

4. Prove $(1.44)_2$.
5. Prove $(1.53)_2$.
6. Prove for nonsingular symmetric tensor $\mathbf{A} = \mathbf{A}^{\mathrm{T}}$:

$$\left(\frac{\partial \mathbf{A}^{-1}}{\partial \mathbf{A}}\right)_{ijkl} = -\frac{1}{2}(A_{ik}^{-1}A_{lj}^{-1} + A_{il}^{-1}A_{kj}^{-1}).$$

(1.91)

7. Prove for scalar φ, vector $\mathbf{a}$, and tensor $\mathbf{B}$:

$$\mathrm{curl\,grad}\varphi = \mathbf{0},$$

$$\mathrm{div\,curl}\mathbf{a} = 0,$$

$$\mathrm{div}(\varphi\mathbf{a}) = \varphi\mathrm{div}\mathbf{a} + \mathbf{a}\cdot\mathrm{grad}\varphi,$$

$$\mathrm{div}(\mathbf{B}^{\mathrm{T}}\mathbf{a}) = (\mathrm{div}\mathbf{B})\cdot\mathbf{a} + \mathbf{B}:\mathrm{grad}\mathbf{a}.$$

(1.92)

8. Prove (1.86).
9. Prove (1.87).
10. Prove (1.88).

References

Itskov M (2013) Tensor algebra and tensor analysis for engineers. Springer, Berlin

Lur'e AI (1990) Nonlinear theory of elasticity. North Holland, Amsterdam

Malvern LE (1969) Introduction to the mechanics of a continuous medium. Prentice-Hall, Englewood Cliffs

Chapter 2
Kinematics

In this chapter we introduce the main concepts of kinematics of continua. These concepts are universal and they do not depend on the choice of specific materials.

2.1 Deformation Gradient

In *continuum mechanics* the atomistic or molecular structure of material is approximated by a continuously distributed set of the so-called *material points* (*material particles*). A continuum material point is an abstraction that is used to designate a small representative volume of real material including many *physical particles* (e.g., atoms, molecules).

Material point that occupied position $\mathbf{x}$ in the *reference configuration* moves to position $\mathbf{y}(\mathbf{x}, t)$ in the *current configuration* of the continuum—Fig. 2.1. It is usually convenient, yet not necessary, to assume that the reference state is the initial one: $\mathbf{x} = \mathbf{y}(\mathbf{x}, 0)$. In accordance with the motion of its material points a body that occupied region Ω_0 with boundary $\partial\Omega_0$ in the initial state moves to region Ω with boundary $\partial\Omega$ in the current state.

If we consider $\mathbf{x}$ as an independent variable then we follow motion of a material point that occupied position $\mathbf{x}$ in the reference configuration. Such description is called *referential* or *material* or *Lagrangean*. If, alternatively, we consider $\mathbf{y}$ as an independent variable then we follow motion of *various* material points passing through the fixed spatial point $\mathbf{y}$ in the current configuration. The latter description is called *spatial* or *Eulerian*. The Eulerian description is often preferable when the evolution of the body boundaries is known in advance like in many problems of fluid mechanics while the Lagrangean description is often preferable when the evolution of the body boundaries is not known in advance like in many problems of solid mechanics. Such a division is conditional, of course, and we will use both Lagrangean and Eulerian descriptions in this book.

© Springer Science+Business Media Singapore 2016
K. Volokh, *Mechanics of Soft Materials*, DOI 10.1007/978-981-10-1599-1_2

Fig. 2.1 Deformation

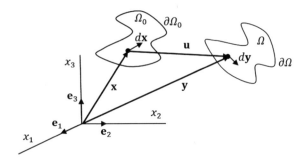

An infinitesimal material fiber at points $\mathbf{x}$ and $\mathbf{y}$ before and after deformation accordingly are related by the linear mapping

$$d\mathbf{y} = \mathbf{F}d\mathbf{x}, \tag{2.1}$$

where[1]

$$\mathbf{F} = \text{Grad}\,\mathbf{y} = \frac{\partial \mathbf{y}}{\partial \mathbf{x}} = \frac{\partial y_i}{\partial x_j}\mathbf{e}_i \otimes \mathbf{e}_j \tag{2.2}$$

is the *deformation gradient*. This tensor is related to two configurations simultaneously and because of that it is called *two-point*.

We can also use the displacement vector, $\mathbf{u} = \mathbf{y} - \mathbf{x}$, to get

$$\mathbf{F} = \text{Grad}(\mathbf{x} + \mathbf{u}) = \mathbf{1} + \mathbf{H}, \tag{2.3}$$

where

$$\mathbf{H} = \text{Grad}\,\mathbf{u} = \frac{\partial u_i}{\partial x_j}\mathbf{e}_i \otimes \mathbf{e}_j \tag{2.4}$$

is the *displacement gradient*.

It is possible to calculate any deformation in the vicinity of a given point when the deformation gradient is known there. We consider deformations of volume, area, and fiber.

We start with the volume deformation—Fig. 2.2

In this case we have

$$d\mathbf{y}^{(m)} = \mathbf{F}d\mathbf{x}^{(m)}, \tag{2.5}$$

---

[1] We capitalize the first character in differential operators: "Grad", "Div", "Curl", when differentiation is with respect to $\mathbf{x}$. The operators are written as usual: "grad", "div", "curl", when differentiation is with respect to $\mathbf{y}$.

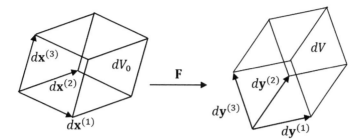

Fig. 2.2 Volume mapping

and, by a direct calculation,

$$
\begin{aligned}
dV &= \det \begin{bmatrix} dy_1^{(1)} & dy_2^{(1)} & dy_3^{(1)} \\ dy_1^{(2)} & dy_2^{(2)} & dy_3^{(2)} \\ dy_1^{(3)} & dy_2^{(3)} & dy_3^{(3)} \end{bmatrix} \\
&= \det \begin{bmatrix} F_{1j}dx_j^{(1)} & F_{2j}dx_j^{(1)} & F_{3j}dx_j^{(1)} \\ F_{1j}dx_j^{(2)} & F_{2j}dx_j^{(2)} & F_{3j}dx_j^{(2)} \\ F_{1j}dx_j^{(3)} & F_{2j}dx_j^{(3)} & F_{3j}dx_j^{(3)} \end{bmatrix} \\
&= \det \begin{bmatrix} dx_1^{(1)} & dx_2^{(1)} & dx_3^{(1)} \\ dx_1^{(2)} & dx_2^{(2)} & dx_3^{(2)} \\ dx_1^{(3)} & dx_2^{(3)} & dx_3^{(3)} \end{bmatrix} \det \begin{bmatrix} F_{11} & F_{21} & F_{31} \\ F_{12} & F_{22} & F_{32} \\ F_{13} & F_{23} & F_{33} \end{bmatrix} \\
&= JdV_0,
\end{aligned}
\tag{2.6}
$$

where

$$
J = \det \mathbf{F} > 0.
\tag{2.7}
$$

The physical meaning of the latter restriction is simple—material cannot disappear during deformation.

In the case of the area deformation—Fig. 2.3—we have for a cylinder built on the infinitesimal base area

$$
\begin{aligned}
dV_0 &= dA_0 \mathbf{n}_0 \cdot d\mathbf{x}, \\
dV &= dA \mathbf{n} \cdot d\mathbf{y} = dA \mathbf{n} \cdot \mathbf{F} d\mathbf{x}.
\end{aligned}
\tag{2.8}
$$

Using (2.6) we derive

$$
dA \mathbf{n} \cdot \mathbf{F} d\mathbf{x} = JdA_0 \mathbf{n}_0 \cdot d\mathbf{x},
\tag{2.9}
$$

and, consequently,

$$
(dA \mathbf{F}^{\mathrm{T}} \mathbf{n} - JdA_0 \mathbf{n}_0) \cdot d\mathbf{x} = 0.
\tag{2.10}
$$

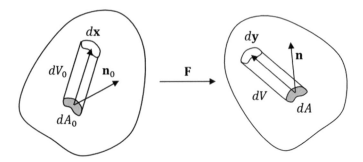

Fig. 2.3 Area mapping

Since $d\mathbf{x}$ is arbitrary we can write down the *Nanson formula*

$$\mathbf{n}dA = J\mathbf{F}^{-\mathrm{T}}\mathbf{n}_0 dA_0. \tag{2.11}$$

Now, we define the fiber stretch—Fig. 2.4—in direction $\mathbf{m}$, $|\mathbf{m}| = 1$,

$$\lambda(\mathbf{m}) = \frac{|d\mathbf{y}|}{|d\mathbf{x}|} = \frac{|\mathbf{F}d\mathbf{x}|}{|d\mathbf{x}|} = |\mathbf{F}\mathbf{m}|. \tag{2.12}$$

We can also define the change of the angle between two fibers—Fig. 2.5—by using stretches as follows, for example,

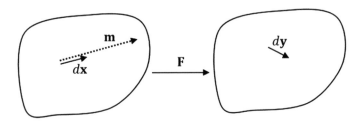

Fig. 2.4 Fiber mapping

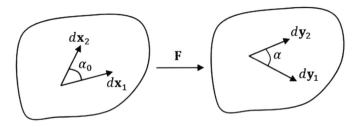

Fig. 2.5 Angle mapping

Fig. 2.6 Simple shear

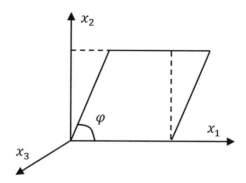

$$\cos \alpha = \frac{d\mathbf{y}_1 \cdot d\mathbf{y}_2}{|d\mathbf{y}_1| |d\mathbf{y}_2|} = \frac{(\mathbf{F}\mathbf{m}_1) \cdot (\mathbf{F}\mathbf{m}_2)}{\lambda(\mathbf{m}_1)\lambda(\mathbf{m}_2)},$$
$$\cos \alpha_0 = \frac{d\mathbf{x}_1 \cdot d\mathbf{x}_2}{|d\mathbf{x}_1| |d\mathbf{x}_2|} = \mathbf{m}_1 \cdot \mathbf{m}_2. \tag{2.13}$$

To illustrate the formulas above we consider the *simple shear* deformation—Fig. 2.6.

We designate the amount of shear by $\gamma = \tan(\pi/2 - \varphi) = \cot \varphi$. The law of motion (deformation) takes form: $y_1 = x_1 + \gamma x_2, y_2 = x_2, y_3 = x_3$. The deformation gradient is

$$\mathbf{F} = \frac{\partial y_i}{\partial x_j} \mathbf{e}_i \otimes \mathbf{e}_j = \mathbf{1} + \gamma \mathbf{e}_1 \otimes \mathbf{e}_2,$$

and we obtain the following stretches for the axial directions

$$\lambda(\mathbf{e}_1) = \sqrt{(\mathbf{F}\mathbf{e}_1) \cdot (\mathbf{F}\mathbf{e}_1)} = \sqrt{\mathbf{e}_1 \cdot \mathbf{e}_1} = 1,$$
$$\lambda(\mathbf{e}_2) = \sqrt{(\mathbf{F}\mathbf{e}_2) \cdot (\mathbf{F}\mathbf{e}_2)} = \sqrt{(\mathbf{e}_2 + \gamma\mathbf{e}_1) \cdot (\mathbf{e}_2 + \gamma\mathbf{e}_1)} = \sqrt{1 + \gamma^2},$$

and the right angle between the directions becomes

$$\alpha = \arccos \frac{(\mathbf{F}\mathbf{e}_1) \cdot (\mathbf{F}\mathbf{e}_2)}{|\mathbf{F}\mathbf{e}_1| |\mathbf{F}\mathbf{e}_2|} = \arccos \frac{\mathbf{e}_1 \cdot (\mathbf{e}_2 + \gamma\mathbf{e}_1)}{\sqrt{1 + \gamma^2}} = \arccos \frac{\gamma}{\sqrt{1 + \gamma^2}}.$$

2.2 Deformation Gradient in Curvilinear Coordinates

In this section we consider the deformation gradient in curvilinear coordinates. To be specific we choose the deformation law in cylindrical coordinates R, Φ, Z before and r, φ, z after the deformation respectively

$$r = r(R, \Phi, Z), \quad \varphi = \varphi(R, \Phi, Z), \quad z = z(R, \Phi, Z). \tag{2.14}$$

We further introduce the natural curvilinear basis vectors for the reference

$$\mathbf{G}_R = \cos\Phi\mathbf{e}_1 + \sin\Phi\mathbf{e}_2, \quad \mathbf{G}_\Phi = -\sin\Phi\mathbf{e}_1 + \cos\Phi\mathbf{e}_2, \quad \mathbf{G}_Z = \mathbf{e}_3, \quad (2.15)$$

and current configurations

$$\mathbf{g}_r = \cos\varphi\mathbf{e}_1 + \sin\varphi\mathbf{e}_2, \quad \mathbf{g}_\varphi = -\sin\varphi\mathbf{e}_1 + \cos\varphi\mathbf{e}_2, \quad \mathbf{g}_z = \mathbf{e}_3, \quad (2.16)$$

accordingly.

Now, the deformation gradient can be written as follows

$$\mathbf{F} = \text{Grad}\mathbf{y} = \frac{\partial\mathbf{y}}{\partial R}\otimes\mathbf{G}_R + \frac{1}{R}\frac{\partial\mathbf{y}}{\partial\Phi}\otimes\mathbf{G}_\Phi + \frac{\partial\mathbf{y}}{\partial Z}\otimes\mathbf{G}_Z, \quad (2.17)$$

where

$$
\begin{aligned}
\mathbf{y} &= y_1\mathbf{e}_1 + y_2\mathbf{e}_2 + y_3\mathbf{e}_3 \\
&= r\cos\varphi(\cos\varphi\mathbf{g}_r - \sin\varphi\mathbf{g}_\varphi) + r\sin\varphi(\sin\varphi\mathbf{g}_r + \cos\varphi\mathbf{g}_\varphi) + z\mathbf{g}_z \\
&= r\mathbf{g}_r + z\mathbf{g}_z.
\end{aligned}
\quad (2.18)
$$

Substituting (2.18) in (2.17) we obtain

$$
\begin{aligned}
\mathbf{F} &= \frac{\partial r}{\partial R}\mathbf{g}_r\otimes\mathbf{G}_R + r\frac{\partial\mathbf{g}_r}{\partial R}\otimes\mathbf{G}_R + \frac{\partial z}{\partial R}\mathbf{g}_z\otimes\mathbf{G}_R \\
&\quad + \frac{1}{R}\frac{\partial r}{\partial\Phi}\mathbf{g}_r\otimes\mathbf{G}_\Phi + \frac{r}{R}\frac{\partial\mathbf{g}_r}{\partial\Phi}\otimes\mathbf{G}_\Phi + \frac{1}{R}\frac{\partial z}{\partial\Phi}\mathbf{g}_z\otimes\mathbf{G}_\Phi \\
&\quad + \frac{\partial r}{\partial Z}\mathbf{g}_r\otimes\mathbf{G}_Z + r\frac{\partial\mathbf{g}_r}{\partial Z}\otimes\mathbf{G}_Z + \frac{\partial z}{\partial Z}\mathbf{g}_z\otimes\mathbf{G}_Z,
\end{aligned}
\quad (2.19)
$$

where

$$
\begin{aligned}
\frac{\partial\mathbf{g}_r}{\partial R} &= \frac{\partial\mathbf{g}_r}{\partial r}\frac{\partial r}{\partial R} + \frac{\partial\mathbf{g}_r}{\partial\varphi}\frac{\partial\varphi}{\partial R} + \frac{\partial\mathbf{g}_r}{\partial z}\frac{\partial z}{\partial R} = \frac{\partial\varphi}{\partial R}\mathbf{g}_\varphi, \\
\frac{\partial\mathbf{g}_r}{\partial\Phi} &= \frac{\partial\mathbf{g}_r}{\partial r}\frac{\partial r}{\partial\Phi} + \frac{\partial\mathbf{g}_r}{\partial\varphi}\frac{\partial\varphi}{\partial\Phi} + \frac{\partial\mathbf{g}_r}{\partial z}\frac{\partial z}{\partial\Phi} = \frac{\partial\varphi}{\partial\Phi}\mathbf{g}_\varphi, \\
\frac{\partial\mathbf{g}_r}{\partial Z} &= \frac{\partial\mathbf{g}_r}{\partial r}\frac{\partial r}{\partial Z} + \frac{\partial\mathbf{g}_r}{\partial\varphi}\frac{\partial\varphi}{\partial Z} + \frac{\partial\mathbf{g}_r}{\partial z}\frac{\partial z}{\partial Z} = \frac{\partial\varphi}{\partial Z}\mathbf{g}_\varphi.
\end{aligned}
\quad (2.20)
$$

After simplifications, we obtain

$$
\begin{aligned}
\mathbf{F} &= \frac{\partial r}{\partial R}\mathbf{g}_r\otimes\mathbf{G}_R + \frac{1}{R}\frac{\partial r}{\partial\Phi}\mathbf{g}_r\otimes\mathbf{G}_\Phi + \frac{\partial r}{\partial Z}\mathbf{g}_r\otimes\mathbf{G}_Z \\
&\quad + r\frac{\partial\varphi}{\partial R}\mathbf{g}_\varphi\otimes\mathbf{G}_R + \frac{r}{R}\frac{\partial\varphi}{\partial\Phi}\mathbf{g}_\varphi\otimes\mathbf{G}_\Phi + r\frac{\partial\varphi}{\partial Z}\mathbf{g}_\varphi\otimes\mathbf{G}_Z \\
&\quad + \frac{\partial z}{\partial R}\mathbf{g}_z\otimes\mathbf{G}_R + \frac{1}{R}\frac{\partial z}{\partial\Phi}\mathbf{g}_z\otimes\mathbf{G}_\Phi + \frac{\partial z}{\partial Z}\mathbf{g}_z\otimes\mathbf{G}_Z.
\end{aligned}
\quad (2.21)
$$

2.3 Polar Decomposition of Deformation Gradient

Let us square the expression for stretch (2.12) and rewrite it as follows

$$\lambda^2(\mathbf{m}) = (\mathbf{Fm}) \cdot (\mathbf{Fm}) = \mathbf{m} \cdot \mathbf{F}^{\mathrm{T}}\mathbf{Fm} = \mathbf{m} \cdot \mathbf{Cm}, \qquad (2.22)$$

where

$$\mathbf{C} = \mathbf{F}^{\mathrm{T}}\mathbf{F} \qquad (2.23)$$

is the *right* Cauchy–Green tensor.

Choosing $\mathbf{m} = \mathbf{m}^{(i)}$ as an eigenvector of tensor $\mathbf{C}$ we have

$$\lambda^2(\mathbf{m}^{(i)}) = \mathbf{m}^{(i)} \cdot \mathbf{Cm}^{(i)} = \mathbf{m}^{(i)} \cdot \zeta_i \mathbf{m}^{(i)} = \zeta_i, \qquad (2.24)$$

where ζ_i is the corresponding eigenvalue of $\mathbf{C}$.

The latter equation means that eigenvalues of the right Cauchy–Green tensor are equal to the squared stretches in principal directions. Thus, we can write the spectral decomposition of $\mathbf{C}$ in the form

$$\mathbf{C} = \lambda_1^2 \mathbf{m}^{(1)} \otimes \mathbf{m}^{(1)} + \lambda_2^2 \mathbf{m}^{(2)} \otimes \mathbf{m}^{(2)} + \lambda_3^2 \mathbf{m}^{(3)} \otimes \mathbf{m}^{(3)}. \qquad (2.25)$$

Now, we define the *right stretch* tensor as a square root of the right Cauchy–Green tensor

$$\mathbf{U} = \sqrt{\mathbf{C}} = \lambda_1 \mathbf{m}^{(1)} \otimes \mathbf{m}^{(1)} + \lambda_2 \mathbf{m}^{(2)} \otimes \mathbf{m}^{(2)} + \lambda_3 \mathbf{m}^{(3)} \otimes \mathbf{m}^{(3)}, \qquad (2.26)$$

where all principal stretches are positive.

We assume then that any deformation gradient can be *multiplicatively* decomposed as

$$\mathbf{F} = \mathbf{RU}. \qquad (2.27)$$

This is called the *polar decomposition* of the deformation gradient and, consequently, we have

$$\mathbf{R} = \mathbf{FU}^{-1}. \qquad (2.28)$$

Let us analyze properties of $\mathbf{R}$. First, we observe that it is *orthogonal*

$$\mathbf{R}^{\mathrm{T}}\mathbf{R} = (\mathbf{FU}^{-1})^{\mathrm{T}}\mathbf{FU}^{-1} = \mathbf{U}^{-\mathrm{T}}\mathbf{F}^{\mathrm{T}}\mathbf{FU}^{-1} = \mathbf{U}^{-\mathrm{T}}\mathbf{U}^2\mathbf{U}^{-1} = \mathbf{1}. \qquad (2.29)$$

Orthogonal tensors do not change lengths of vectors that they map

$$|d\mathbf{y}| = \sqrt{d\mathbf{y} \cdot d\mathbf{y}} = \sqrt{(\mathbf{R}d\mathbf{x}) \cdot (\mathbf{R}d\mathbf{x})} = \sqrt{d\mathbf{x} \cdot \mathbf{R}^{\mathrm{T}}\mathbf{R}d\mathbf{x}} = \sqrt{d\mathbf{x} \cdot d\mathbf{x}} = |d\mathbf{x}|. \qquad (2.30)$$

In addition, we observe (for $\det \mathbf{F} > 0$)

$$\det \mathbf{R} = \frac{\det \mathbf{F}}{\det \mathbf{U}} = \frac{\sqrt{\det \mathbf{F}^T \det \mathbf{F}}}{\det \mathbf{U}} = \frac{\sqrt{\det \mathbf{C}}}{\det \mathbf{U}} = \frac{\sqrt{\det \mathbf{U}^2}}{\det \mathbf{U}} = \frac{\det \mathbf{U}}{\det \mathbf{U}} = 1. \quad (2.31)$$

The latter is the property of the *rotation* tensor. Thus, $\mathbf{R}$ is the *proper orthogonal* or rotation tensor.

Finally we note that the polar decomposition can be interpreted as the successive stretch and rotation—Fig. 2.7.

It is possible, of course, to change the order of stretch and rotation

$$\mathbf{F} = \mathbf{VR}, \quad (2.32)$$

where $\mathbf{V}$ is called the *left stretch* tensor.

By a direct calculation we have

$$\mathbf{V} = \mathbf{FR}^{-1} = \mathbf{FR}^T = \mathbf{RUR}^T = \mathbf{V}^T, \quad (2.33)$$

which means that the left stretch tensor is the rotated right stretch tensor and, consequently, they have the same eigenvalues—principal stretches, while their eigenvectors (principal directions) are different.

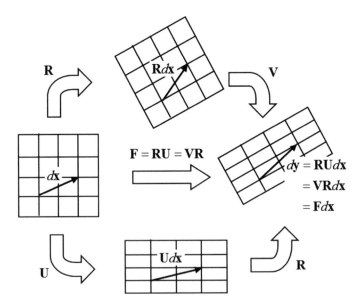

Fig. 2.7 Polar decomposition of deformation gradient

With account of the spectral decomposition of $\mathbf{U}$ we have

$$\mathbf{V} = \lambda_1 \mathbf{n}^{(1)} \otimes \mathbf{n}^{(1)} + \lambda_2 \mathbf{n}^{(2)} \otimes \mathbf{n}^{(2)} + \lambda_3 \mathbf{n}^{(3)} \otimes \mathbf{n}^{(3)}, \tag{2.34}$$

where (no sum on i)

$$\mathbf{n}^{(i)} \otimes \mathbf{n}^{(i)} = \mathbf{R}(\mathbf{m}^{(i)} \otimes \mathbf{m}^{(i)})\mathbf{R}^{\mathrm{T}} = \mathbf{R}\mathbf{m}^{(i)} \otimes \mathbf{R}\mathbf{m}^{(i)}. \tag{2.35}$$

To clarify the meaning of the principal directions of $\mathbf{V}$ we square the tensor as follows

$$\mathbf{V}^2 = (\mathbf{R}\mathbf{U}\mathbf{R}^{\mathrm{T}})(\mathbf{R}\mathbf{U}\mathbf{R}^{\mathrm{T}}) = \mathbf{R}\mathbf{U}\mathbf{U}\mathbf{R}^{\mathrm{T}} = \mathbf{R}\mathbf{U}(\mathbf{R}\mathbf{U})^{\mathrm{T}} = \mathbf{F}\mathbf{F}^{\mathrm{T}} = \mathbf{B}, \tag{2.36}$$

and, consequently,

$$\mathbf{B} = \lambda_1^2 \mathbf{n}^{(1)} \otimes \mathbf{n}^{(1)} + \lambda_2^2 \mathbf{n}^{(2)} \otimes \mathbf{n}^{(2)} + \lambda_3^2 \mathbf{n}^{(3)} \otimes \mathbf{n}^{(3)}, \tag{2.37}$$

where $\mathbf{B}$ is the *left* Cauchy–Green tensor (also called Finger strain tensor), which principal directions coincide with the principal directions of $\mathbf{V}$ while the principal values of $\mathbf{B}$ are squared principal stretches.

Unfortunately, we cannot directly write the relations between the directions of eigenvectors $\mathbf{m}^{(i)}$ and $\mathbf{n}^{(i)}$ in the reference and current configurations because these directions are not defined uniquely and can always be reversed. However, we can *define* the principal directions uniquely by the following procedure. Assume, for example, that the principal directions in the reference configuration, $\mathbf{m}^{(i)}$, are uniquely chosen then we calculate the principal directions in the current configuration as follows

$$\mathbf{n}^{(i)} = \mathbf{R}\mathbf{m}^{(i)}. \tag{2.38}$$

Of course, we could start with the current configuration otherwise.

Finally, we can calculate the spectral decomposition, which is the *singular value decomposition*, of the deformation gradient as follows

$$\mathbf{F} = \mathbf{R}\mathbf{U} = \lambda_1 \mathbf{R}\mathbf{m}^{(1)} \otimes \mathbf{m}^{(1)} + \lambda_2 \mathbf{R}\mathbf{m}^{(2)} \otimes \mathbf{m}^{(2)} + \lambda_3 \mathbf{R}\mathbf{m}^{(3)} \otimes \mathbf{m}^{(3)}$$

$$= \lambda_1 \mathbf{n}^{(1)} \otimes \mathbf{m}^{(1)} + \lambda_2 \mathbf{n}^{(2)} \otimes \mathbf{m}^{(2)} + \lambda_3 \mathbf{n}^{(3)} \otimes \mathbf{m}^{(3)}. \tag{2.39}$$

A nice analytical example on the spectral and polar decompositions of the deformation gradient was found by Marsden and Hughes (1983). They considered the following law of deformation: $y_1 = \sqrt{3}x_1 + x_2$, $y_2 = 2x_2$, $y_3 = x_3$; and they calculated the corresponding quantities in Cartesian coordinates

$$[\mathbf{F}] = \begin{bmatrix} \sqrt{3} & 1 & 0 \\ 0 & 2 & 0 \\ 0 & 0 & 1 \end{bmatrix}, \quad [\mathbf{C}] = \begin{bmatrix} 3 & \sqrt{3} & 0 \\ \sqrt{3} & 5 & 0 \\ 0 & 0 & 1 \end{bmatrix},$$

$$\lambda_1 = \sqrt{6}, \ \left[\mathbf{m}^{(1)}\right] = \frac{1}{2}\begin{bmatrix} 1 \\ \sqrt{3} \\ 0 \end{bmatrix}, \quad \lambda_2 = \sqrt{2}, \ \left[\mathbf{m}^{(2)}\right] = \frac{1}{2}\begin{bmatrix} \sqrt{3} \\ -1 \\ 0 \end{bmatrix}, \quad \lambda_3 = 1, \ \left[\mathbf{m}^{(3)}\right] = \begin{bmatrix} 0 \\ 0 \\ 1 \end{bmatrix},$$

$$[\mathbf{U}] = \frac{1}{2\sqrt{2}}\begin{bmatrix} 3+\sqrt{3} & 3-\sqrt{3} & 0 \\ 3-\sqrt{3} & 1+3\sqrt{3} & 0 \\ 0 & 0 & 2\sqrt{2} \end{bmatrix}, \quad \left[\mathbf{U}^{-1}\right] = \frac{1}{4\sqrt{6}}\begin{bmatrix} 1+3\sqrt{3} & \sqrt{3}-3 & 0 \\ \sqrt{3}-3 & 3+\sqrt{3} & 0 \\ 0 & 0 & 4\sqrt{6} \end{bmatrix},$$

$$[\mathbf{R}] = \frac{1}{2\sqrt{2}}\begin{bmatrix} 1+\sqrt{3} & \sqrt{3}-1 & 0 \\ 1-\sqrt{3} & 1+\sqrt{3} & 0 \\ 0 & 0 & 2\sqrt{2} \end{bmatrix}, \quad [\mathbf{V}] = \frac{1}{\sqrt{2}}\begin{bmatrix} 1+\sqrt{3} & \sqrt{3}-1 & 0 \\ \sqrt{3}-1 & 1+\sqrt{3} & 0 \\ 0 & 0 & \sqrt{2} \end{bmatrix}.$$

2.4   Strain

Strain is a geometric measure of deformation and it can be introduced in various ways. We start with one-dimensional measures for the change of the length of a material fiber—Fig. 2.8.

We can introduce the *engineering strain*, *logarithmic strain*, or the *Green strain* accordingly

$$E_E = \frac{L - L_0}{L_0} = \lambda - 1,$$

$$E_L = \int_{L_0}^{L} \frac{dL}{L_0} = \ln\frac{L}{L_0} = \ln\lambda, \qquad (2.40)$$

$$E_G = \frac{L^2 - L_0^2}{2L_0^2} = \frac{1}{2}(\lambda^2 - 1).$$

In order to generalize one-dimensional strains to the three-dimensional ones we assume that the previous formulas are valid in the principal directions of the reference configuration. In this case, the three-dimensional strain tensors take forms

$$\mathbf{E}_E = \sum_{i=1}^{3}(\lambda_i - 1)\mathbf{m}^{(i)} \otimes \mathbf{m}^{(i)} = \mathbf{U} - \mathbf{1},$$

$$\mathbf{E}_L = \sum_{i=1}^{3}(\ln\lambda_i)\mathbf{m}^{(i)} \otimes \mathbf{m}^{(i)} = \ln\mathbf{U}, \qquad (2.41)$$

$$\mathbf{E}_G = \sum_{i=1}^{3}\frac{1}{2}(\lambda_i^2 - 1)\mathbf{m}^{(i)} \otimes \mathbf{m}^{(i)} = \frac{1}{2}(\mathbf{U}^2 - \mathbf{1}).$$

Fig. 2.8   Strain

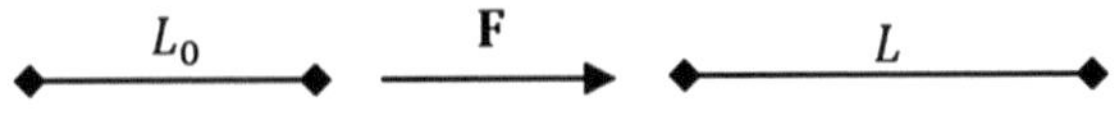

The Green strain tensor is very popular and it can be written without the suffix

$$\mathbf{E} = \frac{1}{2}(\mathbf{U}^2 - \mathbf{1}) = \frac{1}{2}(\mathbf{C} - \mathbf{1}) = \frac{1}{2}(\mathbf{F}^T\mathbf{F} - \mathbf{1}). \tag{2.42}$$

2.5 Motion

Velocity and acceleration vectors are defined as material time derivatives of the placement vector $\mathbf{y}(\mathbf{x}, t)$ as follows

$$\mathbf{v} = \frac{d\mathbf{y}(\mathbf{x}, t)}{dt} = \dot{\mathbf{y}} = \dot{\mathbf{x}} + \dot{\mathbf{u}} = \dot{\mathbf{u}},$$
$$\mathbf{a} = \frac{d\mathbf{v}}{dt} = \dot{\mathbf{v}}. \tag{2.43}$$

When the Eulerian or spatial description is used it is necessary to apply the chain rule for differentiation of function $f(\mathbf{y}(t), t)$

$$\frac{df}{dt} = \dot{f}(\mathbf{y}(t), t) = \frac{\partial f}{\partial t} + \text{grad}f \cdot \frac{\partial \mathbf{y}}{\partial t} = \frac{\partial f}{\partial t} + \text{grad}f \cdot \mathbf{v}. \tag{2.44}$$

For example, we have for the acceleration vector

$$\mathbf{a} = \frac{d\mathbf{v}}{dt} = \frac{\partial \mathbf{v}}{\partial t} + \mathbf{L}\mathbf{v}, \tag{2.45}$$

in which another important quantity—*velocity gradient*—is introduced

$$\mathbf{L} = \text{grad}\mathbf{v} = \frac{\partial \mathbf{v}}{\partial \mathbf{y}}. \tag{2.46}$$

We emphasize that the partial time derivatives are taken for $\mathbf{y}$ fixed

$$\frac{\partial f}{\partial t} \equiv \left(\frac{\partial f(\mathbf{y}, t)}{\partial t}\right)_{\mathbf{y} \text{ fixed}}. \tag{2.47}$$

Another way to calculate the velocity gradient comes from identity

$$\dot{\mathbf{F}} = \frac{d}{dt}\text{Grad}\mathbf{y} = \text{Grad}\dot{\mathbf{y}} = \text{Grad}\mathbf{v} = (\text{grad}\mathbf{v})\mathbf{F} = \mathbf{L}\mathbf{F}. \tag{2.48}$$

From the latter equation we get

$$\mathbf{L} = \dot{\mathbf{F}}\mathbf{F}^{-1}. \tag{2.49}$$

The velocity gradient can be decomposed into symmetric and skew parts

$$\mathbf{L} = \mathbf{D} + \mathbf{W}, \quad \mathbf{D} = \frac{1}{2}(\mathbf{L} + \mathbf{L}^{\mathrm{T}}), \quad \mathbf{W} = \frac{1}{2}(\mathbf{L} - \mathbf{L}^{\mathrm{T}}), \tag{2.50}$$

where $\mathbf{D}$ and $\mathbf{W}$ are the *deformation rate* and the *spin (vorticity)* tensors accordingly.

2.6   Rigid Body Motion

The *rigid body motion* (RBM) superimposed on the current configuration—Fig. 2.9—is of importance for constitutive modeling, which will be discussed in the coming chapters. It is generally required that the constitutive laws should not be affected by the superimposed rigid body motion—they should be *objective*.

The superimposed RBM, designated with asterisk, can be described as follows

$$\mathbf{y}^* = \mathbf{Q}(t)\mathbf{y} + \mathbf{c}(t), \tag{2.51}$$

where

$$\mathbf{Q}^{\mathrm{T}} = \mathbf{Q}^{-1}, \quad \det \mathbf{Q} = 1 \tag{2.52}$$

is a proper-orthogonal tensor.

We remind the reader that since $\det \mathbf{Q} = 1 > 0$ then material does not disappear and tensor $\mathbf{Q}$ describes rotation.

The transformation law for a material fiber takes form

$$\mathbf{s}^* = \mathbf{y}_2^* - \mathbf{y}_1^* = \mathbf{Q}(\mathbf{y}_2 - \mathbf{y}_1) = \mathbf{Q}\mathbf{s}. \tag{2.53}$$

This motion preserves length

$$\left|\mathbf{s}^*\right| = \sqrt{\mathbf{s}^* \cdot \mathbf{s}^*} = \sqrt{\mathbf{s} \cdot \mathbf{Q}^{\mathrm{T}}\mathbf{Q}\mathbf{s}} = \sqrt{\mathbf{s} \cdot \mathbf{1}\mathbf{s}} = |\mathbf{s}|. \tag{2.54}$$

Besides, it preserves angles between fibers. Check it.

All quantities related to the reference configuration at $t = 0$ are unaffected by RBM and only quantities related to the current configuration are affected by RBM.

Fig. 2.9 Superimposed rigid body motion

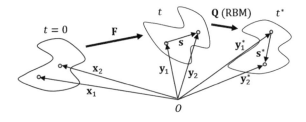

Vector quantities that transform under superimposed RBM following the rule

$$\mathbf{s}^* = \mathbf{Qs} \tag{2.55}$$

are called objective, i.e. unaffected by RBM.

Not all vectors are objective. For example, velocity and acceleration vectors are not objective

$$\mathbf{v}^* = \dot{\mathbf{y}}^* = \frac{d}{dt}(\mathbf{Qy} + \mathbf{c}) = \mathbf{Q}\dot{\mathbf{y}} + \dot{\mathbf{Q}}\mathbf{y} + \dot{\mathbf{c}} = \mathbf{Qv} + \mathbf{\Omega}(\mathbf{y}^* - \mathbf{c}) + \dot{\mathbf{c}},$$
$$\mathbf{a}^* = \dot{\mathbf{v}}^* = \mathbf{Qa} + \ddot{\mathbf{c}} + (\dot{\mathbf{\Omega}} - \mathbf{\Omega}^2)(\mathbf{y}^* - \mathbf{c}) + 2\mathbf{\Omega}(\mathbf{v}^* - \dot{\mathbf{c}}), \tag{2.56}$$

where

$$\mathbf{y} = \mathbf{Q}(\mathbf{y}^* - \mathbf{c}),$$
$$\mathbf{\Omega} = \dot{\mathbf{Q}}\mathbf{Q}^{-1} = \dot{\mathbf{Q}}\mathbf{Q}^{\mathrm{T}} = -\mathbf{\Omega}^{\mathrm{T}}, \tag{2.57}$$
$$\frac{d}{dt}(\mathbf{QQ}^{\mathrm{T}}) = \dot{\mathbf{Q}}\mathbf{Q}^{\mathrm{T}} + \mathbf{Q}\dot{\mathbf{Q}}^{\mathrm{T}} = \mathbf{0},$$

and tensor $\mathbf{\Omega}$ is the spin of RBM.

Second-order tensors defined in the current configuration are called objective if they preserve objectivity of the vectors that they map. Let tensor $\mathbf{A}$ map objective vector $\mathbf{s}$ into objective vector $\mathbf{r}$:

$$\mathbf{r} = \mathbf{As}. \tag{2.58}$$

Then, tensor $\mathbf{A}^*$ maps proper transformations of the objective vectors

$$\mathbf{r}^* = \mathbf{A}^*\mathbf{s}^*, \tag{2.59}$$

or

$$\mathbf{Qr} = \mathbf{A}^*\mathbf{Qs}, \tag{2.60}$$

and, consequently, we have

$$\mathbf{r} = \mathbf{Q}^{\mathrm{T}}\mathbf{A}^*\mathbf{Qs}. \tag{2.61}$$

Comparing (2.58) and (2.61) and assuming that the choice of vectors is arbitrary, we obtain the transformation rule for the objective second-order tensor

$$\mathbf{A}^* = \mathbf{QAQ}^{\mathrm{T}}. \tag{2.62}$$

Not all tensors are objective. Let us examine objectivity of the velocity gradient tensor

$$\begin{aligned}
\mathbf{L}^* &= \dot{\mathbf{F}}^*\mathbf{F}^{*-1} \\
&= (\dot{\mathbf{Q}}\mathbf{F} + \mathbf{Q}\dot{\mathbf{F}})\mathbf{F}^{-1}\mathbf{Q}^{-1} \\
&= \dot{\mathbf{Q}}\mathbf{Q}^{-1} + \mathbf{Q}\dot{\mathbf{F}}\mathbf{F}^{-1}\mathbf{Q}^{-1} \\
&= \mathbf{\Omega} + \mathbf{Q}\mathbf{L}\mathbf{Q}^{\mathrm{T}}.
\end{aligned} \tag{2.63}$$

This tensor is affected by the superimposed RBM and it is not objective. However, its symmetric part—the deformation rate tensor—is objective

$$\begin{aligned}
\mathbf{D}^* &= \frac{1}{2}(\mathbf{L}^* + \mathbf{L}^{*\mathrm{T}}) \\
&= \frac{1}{2}(\mathbf{\Omega} + \mathbf{Q}\mathbf{L}\mathbf{Q}^{\mathrm{T}} + \mathbf{\Omega}^{\mathrm{T}} + \mathbf{Q}\mathbf{L}^{\mathrm{T}}\mathbf{Q}^{\mathrm{T}}) \\
&= \frac{1}{2}\mathbf{Q}(\mathbf{L} + \mathbf{L}^{\mathrm{T}})\mathbf{Q}^{\mathrm{T}} \\
&= \mathbf{Q}\mathbf{D}\mathbf{Q}^{\mathrm{T}}.
\end{aligned} \tag{2.64}$$

We note that the rate of an objective second-order tensor is not objective

$$\dot{\mathbf{A}}^* = \frac{d}{dt}(\mathbf{Q}\mathbf{A}\mathbf{Q}^{\mathrm{T}}) = \dot{\mathbf{Q}}\mathbf{A}\mathbf{Q}^{\mathrm{T}} + \mathbf{Q}\dot{\mathbf{A}}\mathbf{Q}^{\mathrm{T}} + \mathbf{Q}\mathbf{A}\dot{\mathbf{Q}}^{\mathrm{T}}. \tag{2.65}$$

The latter observation triggered various proposals for an objective rate of an objective tensor. For example, we mention the Jaumann-Zaremba, Truesdell, and Oldroyd objective rates respectively

$$\begin{aligned}
\overset{\bullet}{\mathbf{A}} &= \dot{\mathbf{A}} - \mathbf{W}\mathbf{A} - \mathbf{A}\mathbf{W}^{\mathrm{T}}, \\
\overset{\circ}{\mathbf{A}} &= \dot{\mathbf{A}} - \mathbf{L}\mathbf{A} - \mathbf{A}\mathbf{L}^{\mathrm{T}} + (\mathrm{tr}\mathbf{L})\mathbf{A}, \\
\overset{\diamond}{\mathbf{A}} &= \dot{\mathbf{A}} - \mathbf{L}\mathbf{A} - \mathbf{A}\mathbf{L}^{\mathrm{T}}.
\end{aligned} \tag{2.66}$$

The proof of the objectivity of the Oldroyd rate, for example, is by the direct calculation

$$\begin{aligned}
\overset{\diamond}{\mathbf{A}}^* &= \dot{\mathbf{A}}^* - \mathbf{L}^*\mathbf{A}^* - \mathbf{A}^*\mathbf{L}^{*\mathrm{T}} \\
&= \dot{\mathbf{Q}}\mathbf{A}\mathbf{Q}^{\mathrm{T}} + \mathbf{Q}\dot{\mathbf{A}}\mathbf{Q}^{\mathrm{T}} + \mathbf{Q}\mathbf{A}\dot{\mathbf{Q}}^{\mathrm{T}} - (\mathbf{Q}\mathbf{L}\mathbf{Q}^{\mathrm{T}} + \dot{\mathbf{Q}}\mathbf{Q}^{\mathrm{T}})\mathbf{Q}\mathbf{A}\mathbf{Q}^{\mathrm{T}} - \mathbf{Q}\mathbf{A}\mathbf{Q}^{\mathrm{T}}(\mathbf{Q}\mathbf{L}\mathbf{Q}^{\mathrm{T}} + \dot{\mathbf{Q}}\mathbf{Q}^{\mathrm{T}})^{\mathrm{T}} \\
&= \dot{\mathbf{Q}}\mathbf{A}\mathbf{Q}^{\mathrm{T}} + \mathbf{Q}\dot{\mathbf{A}}\mathbf{Q}^{\mathrm{T}} + \mathbf{Q}\mathbf{A}\dot{\mathbf{Q}}^{\mathrm{T}} - \mathbf{Q}\mathbf{L}\mathbf{A}\mathbf{Q}^{\mathrm{T}} - \dot{\mathbf{Q}}\mathbf{A}\mathbf{Q}^{\mathrm{T}} - \mathbf{Q}\mathbf{A}\mathbf{L}^{\mathrm{T}}\mathbf{Q}^{\mathrm{T}} - \mathbf{Q}\mathbf{A}\dot{\mathbf{Q}}^{\mathrm{T}} \\
&= \mathbf{Q}\dot{\mathbf{A}}\mathbf{Q}^{\mathrm{T}} - \mathbf{Q}\mathbf{L}\mathbf{A}\mathbf{Q}^{\mathrm{T}} - \mathbf{Q}\mathbf{A}\mathbf{L}^{\mathrm{T}}\mathbf{Q}^{\mathrm{T}} \\
&= \mathbf{Q}(\dot{\mathbf{A}} - \mathbf{L}\mathbf{A} - \mathbf{A}\mathbf{L}^{\mathrm{T}})\mathbf{Q}^{\mathrm{T}} \\
&= \mathbf{Q}\overset{\diamond}{\mathbf{A}}\mathbf{Q}^{\mathrm{T}}.
\end{aligned}$$

2.7   Lagrangean, Eulerian and Two-Point Tensors

Tensor fields considered in the previous sections can be classified as Lagrangean, Eulerian, and two-point.

Lagrangean tensors are defined on the initial or referential configuration. For example, the right Cauchy–Green tensor $\mathbf{C}$, right stretch tensor $\mathbf{U}$, strain tensors $\mathbf{E}_G$, $\mathbf{E}_E$, $\mathbf{E}_L$ are Lagrangean.

Eulerian tensors are defined on the current configuration. For example, the left Cauchy–Green tensor $\mathbf{B}$, left stretch tensor $\mathbf{V}$, velocity gradient $\mathbf{L}$, deformation rate $\mathbf{D}$, spin $\mathbf{W}$ are Eulerian.

Two-point tensors belong to both initial and current configurations simultaneously. For example, deformation gradient $\mathbf{F}$ and rotation tensor $\mathbf{R}$ are two-point.

Vectors cannot be two-point—they are Eulerian, like $\mathbf{n}^{(i)}$, or Lagrangean, like $\mathbf{m}^{(j)}$.

It is important to follow the character of the tensor (Lagrangean, Eulerian, and two-point) in order to have physically consistent formulations. It is also important to not confuse Eulerian and Lagrangean tensors with the Eulerian and Lagrangean descriptions of motion. Lagrangean description of motion can be used for Eulerian tensors and Eulerian description of motion can be used for Lagrangean tensors.

2.8   Exercises

1. Find principal directions and stretches for the following deformation law

$$y_1 = (1 + \alpha)x_1 + \alpha x_2, \ \ y_2 = -\alpha x_1 + (1 + \alpha)x_2, \ \ y_3 = x_3, \qquad (2.67)$$

   where $\alpha = $ constant.
2. Calculate the polar decomposition of the deformation gradient for the deformation law presented in (2.67).
3. Calculate the Cartesian components of the Green strain for the deformation law presented in (2.67).
4. Derive (2.56).
5. Prove objectivity of the Jaumann-Zaremba and Truesdell rates $(2.66)_{1,2}$.

References

Marsden JE, Hughes TJR (1983) Mathematical foundations of elasticity. Prentice-Hall, Upper Saddle River
Ogden RW (1997) Non-Linear elastic deformations. Dover, Illinois

Chapter 3
Balance Laws

In this chapter we introduce mass and momenta balance laws. Additional balance laws are considered for coupled problems in the subsequent chapters. Generally, balance laws result from our observations in Nature and they have experimental roots which are not discussed here.

3.1 Material Time Derivative of Integral

For the field quantity $\psi(\mathbf{y}(t), t)$ defined in a moving region $\Omega(t)$, whose configuration depends on time t, we calculate the following formula (considering the integral as an infinite sum)

$$
\begin{aligned}
\frac{d}{dt} \int \psi dV(t) &= \int \frac{d}{dt} \{\psi dy_1(t)dy_2(t)dy_3(t)\} \\
&= \int \left\{ \frac{d\psi}{dt} dy_1(t)dy_2(t)dy_3(t) + \psi dv_1(t)dy_2(t)dy_3(t) \right. \\
&\quad \left. + \psi dy_1(t)dv_2(t)dy_3(t) + \psi dy_1(t)dy_2(t)dv_3(t) \right\} \\
&= \int \left\{ \frac{d\psi}{dt} + \psi \frac{\partial v_1}{\partial y_1} + \psi \frac{\partial v_2}{\partial y_2} + \psi \frac{\partial v_3}{\partial y_3} \right\} dV \\
&= \int \left\{ \frac{d\psi}{dt} + \psi \mathrm{div}\mathbf{v} \right\} dV \\
&= \int \left\{ \frac{\partial \psi}{\partial t} + \mathrm{div}(\psi \mathbf{v}) \right\} dV.
\end{aligned}
\tag{3.1}
$$

© Springer Science+Business Media Singapore 2016
K. Volokh, *Mechanics of Soft Materials*, DOI 10.1007/978-981-10-1599-1_3

The last equality is obtained as follows

$$
\begin{aligned}
\frac{d\psi}{dt} + \psi \mathrm{div}\mathbf{v} &= \frac{\partial \psi}{\partial t} + \frac{\partial \psi}{\partial y_i}\frac{\partial y_i}{\partial t} + \psi \frac{\partial v_i}{\partial y_i} \\
&= \frac{\partial \psi}{\partial t} + \frac{\partial \psi}{\partial y_i} v_i + \psi \frac{\partial v_i}{\partial y_i} \\
&= \frac{\partial \psi}{\partial t} + \frac{\partial (\psi v_i)}{\partial y_i} \\
&= \frac{\partial \psi}{\partial t} + \mathrm{div}(\psi \mathbf{v}).
\end{aligned}
\tag{3.2}
$$

3.2 Mass Conservation

The law of mass conservation states that the mass of a body or *any part* of it does not change in the process of deformation. This statement can be formalized as follows

$$
m = \int \rho \, dV = \text{constant},
\tag{3.3}
$$

where ρ is the *mass density* and the integration is over the whole body or any part of it.

Differentiating this formula with respect to time and using (3.1) we obtain

$$
\frac{dm}{dt} = \frac{d}{dt} \int \rho \, dV = \int \left(\frac{d\rho}{dt} + \rho \mathrm{div}\mathbf{v} \right) dV = 0,
\tag{3.4}
$$

Since the equality is obeyed for any part of the body we can *localize* it for an infinitesimal volume

$$
\frac{d\rho}{dt} + \rho \mathrm{div}\mathbf{v} = \frac{\partial \rho}{\partial t} + \mathrm{div}(\rho \mathbf{v}) = 0.
\tag{3.5}
$$

Here, we state *mass conservation* which is a special case of *mass balance* where neither volume nor surface supply of mass is assumed. This conservation "law" is violated in the case of biological tissues, for example. We discuss the general form of balance laws in Sect. 3.5.

3.3 Balance of Linear Momentum

In order to motivate the law of the linear momentum balance in the case of continuum, we start with the balance of linear momentum for a volumeless particle—Newton's law

$$\frac{d}{dt}(m\mathbf{v}) = \mathbf{f}, \tag{3.6}$$

where $m\mathbf{v}$ is the *linear momentum* and $\mathbf{f}$ is the force resultant.

By analogy with Newton's law Euler considered the balance of the linear momentum for body Ω bounded by surface $\partial\Omega$

$$\frac{d}{dt}\int \rho\mathbf{v}dV = \int \mathbf{b}dV + \oint \mathbf{t}dA. \tag{3.7}$$

Here $\mathbf{b}$ is the *body force* per unit current volume and $\mathbf{t}$ is the *surface force* or *traction* per unit current area.

Thus the force resultant in the case of a volumeless particle splits into body and surface forces in the case of continuum.

Let us localize the Euler law. First, differentiating the left-hand side of (3.7) by using (3.1) we get

$$\frac{d}{dt}\int \rho\mathbf{v}dV = \int \left(\frac{d(\rho\mathbf{v})}{dt} + \rho\mathbf{v}\mathrm{divv}\right)dV. \tag{3.8}$$

Then, we rewrite the Euler law in the form

$$\int \mathbf{b}_g dV = \oint \mathbf{t}dA, \tag{3.9}$$

where

$$\mathbf{b}_g = -\mathbf{b} + \frac{d(\rho\mathbf{v})}{dt} + \rho\mathbf{v}\mathrm{divv} \tag{3.10}$$

is the generalized body force.

Now it is necessary to transform the area integral into the volume integral. This is possible via the *Cauchy assumption*—Fig. 3.1.

The body cross-section is defined by the normal vector $\mathbf{n}$ at a chosen material point $\mathbf{y}$ where the force resultant $\triangle\mathbf{t}$ over small area $\triangle A$ is applied. When the area shrinks to zero it is possible to define *stress vector* $\mathbf{t}$ as a force per unit area. It is reasonable to assume, following Cauchy, that the stress vector depends on the material point at $\mathbf{y}$ and the direction of the cross-section $\mathbf{n}$:

$$\mathbf{t} = \lim_{\triangle A \to 0}\frac{\triangle\mathbf{t}}{\triangle A} = \mathbf{t}(\mathbf{y}, \mathbf{n}). \tag{3.11}$$

This simple and intuitively appealing assumption leads to two not very evident corollaries considered in the following subsections.

Fig. 3.1 Cauchy assumption

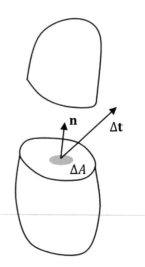

3.3.1 Action and Counteraction

The first corollary of the Cauchy assumption is the third Newton law of *action and counteraction.*

Indeed, let us consider a body partitioned into two parts with the help of surface A^*—Fig. 3.2.

The Euler law for every part of the body is written as follows

$$\int \mathbf{b}_g dV_1 = \int \mathbf{t} dA_1 + \int \mathbf{t}(\mathbf{n}) dA^*,$$
$$\int \mathbf{b}_g dV_2 = \int \mathbf{t} dA_2 + \int \mathbf{t}(-\mathbf{n}) dA^*. \tag{3.12}$$

The sum of these equations reads

$$\int \mathbf{b}_g dV = \oint \mathbf{t} dA + \int [\mathbf{t}(\mathbf{n}) + \mathbf{t}(-\mathbf{n})] dA^*, \tag{3.13}$$

Fig. 3.2 Action and counteraction

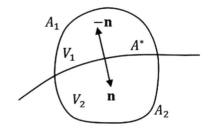

and it implies

$$\int [\mathbf{t}(\mathbf{n}) + \mathbf{t}(-\mathbf{n})]dA^* = \mathbf{0}. \tag{3.14}$$

The latter equality is correct for any surface and, consequently, we can localize it and get the third Newton law of action and counteraction

$$\mathbf{t}(\mathbf{n}) = -\mathbf{t}(-\mathbf{n}). \tag{3.15}$$

3.3.2 Cauchy Stress

The second corollary of the Cauchy assumption is related to the Cauchy stress tensor. Let us define a tetrahedron of height h in direction $\mathbf{n}$ at point $\mathbf{y}$—Fig. 3.3.

The direction cosines allow us to calculate the following areas of the tetrahedron

$$CDB = A, \ COB = An_1, \ COD = An_2, \ DOB = An_3. \tag{3.16}$$

We apply the linear momentum balance to the tetrahedron as follows

$$\int \mathbf{b}_g dV = \int_{CDB} \mathbf{t}(\mathbf{n})dA + \int_{COB} \mathbf{t}(-\mathbf{e}_1)dA + \int_{COD} \mathbf{t}(-\mathbf{e}_2)dA + \int_{DOB} \mathbf{t}(-\mathbf{e}_3)dA. \tag{3.17}$$

According to the mean value theorem and by using (3.16) we have

$$\bar{\mathbf{b}}_g \frac{hA}{6} = \bar{\mathbf{t}}(\mathbf{n})A + \bar{\mathbf{t}}(-\mathbf{e}_1)An_1 + \bar{\mathbf{t}}(-\mathbf{e}_2)An_2 + \bar{\mathbf{t}}(-\mathbf{e}_3)An_3, \tag{3.18}$$

where the barred quantities are calculated inside the proper volume or area.

Setting $h \to 0$ we obtain for the tetrahedron shrinked to the point

$$\mathbf{0} = \mathbf{t}(\mathbf{n}) + \mathbf{t}(-\mathbf{e}_1)n_1 + \mathbf{t}(-\mathbf{e}_2)n_2 + \mathbf{t}(-\mathbf{e}_3)n_3. \tag{3.19}$$

Fig. 3.3 Cauchy tetrahedron

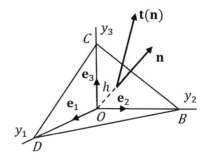

Further, accounting for $n_i = \mathbf{e}_i \cdot \mathbf{n}$ and the law of action and counteraction, $\mathbf{t}(-\mathbf{e}_j) = -\mathbf{t}(\mathbf{e}_j)$, we can write down

$$
\begin{aligned}
\mathbf{t}(\mathbf{n}) &= \mathbf{t}(\mathbf{e}_1)(\mathbf{e}_1 \cdot \mathbf{n}) + \mathbf{t}(\mathbf{e}_2)(\mathbf{e}_2 \cdot \mathbf{n}) + \mathbf{t}(\mathbf{e}_3)(\mathbf{e}_3 \cdot \mathbf{n}) \\
&= (\mathbf{t}(\mathbf{e}_1) \otimes \mathbf{e}_1 + \mathbf{t}(\mathbf{e}_2) \otimes \mathbf{e}_2 + \mathbf{t}(\mathbf{e}_3) \otimes \mathbf{e}_3)\mathbf{n} \\
&= \boldsymbol{\sigma}\mathbf{n}.
\end{aligned}
\tag{3.20}
$$

Here, we introduce the *Cauchy stress tensor* through the tensor products of traction and basis vectors

$$
\boldsymbol{\sigma} = \mathbf{t}(\mathbf{e}_1) \otimes \mathbf{e}_1 + \mathbf{t}(\mathbf{e}_2) \otimes \mathbf{e}_2 + \mathbf{t}(\mathbf{e}_3) \otimes \mathbf{e}_3 = \mathbf{t}(\mathbf{e}_s) \otimes \mathbf{e}_s.
\tag{3.21}
$$

Cartesian components (1.24) of the Cauchy stress tensor can be readily calculated as follows

$$
\begin{aligned}
\sigma_{ij} &= \mathbf{e}_i \otimes \mathbf{e}_j : \boldsymbol{\sigma} \\
&= \mathbf{e}_i \otimes \mathbf{e}_j : \mathbf{t}(\mathbf{e}_s) \otimes \mathbf{e}_s \\
&= (\mathbf{e}_i \cdot \mathbf{t}(\mathbf{e}_s))(\mathbf{e}_j \cdot \mathbf{e}_s) \\
&= \mathbf{e}_i \cdot \mathbf{t}(\mathbf{e}_s)\delta_{js} \\
&= \mathbf{e}_i \cdot \mathbf{t}(\mathbf{e}_j).
\end{aligned}
\tag{3.22}
$$

For example, we have the following stress components for the area with normal $\mathbf{e}_2$—Fig. 3.4

$$
\sigma_{12} = \mathbf{e}_1 \cdot \mathbf{t}(\mathbf{e}_2), \quad \sigma_{22} = \mathbf{e}_2 \cdot \mathbf{t}(\mathbf{e}_2), \quad \sigma_{32} = \mathbf{e}_3 \cdot \mathbf{t}(\mathbf{e}_2),
$$

which means that the components of the Cauchy stress tensor are projections of the stress vector onto the axes of Cartesian coordinates.

The notational convention for the stress components is shown in Fig. 3.5. The first index gives direction of the stress and the second index gives direction of the normal to the area where the stress is applied.

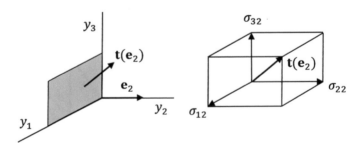

Fig. 3.4 Stress components

Fig. 3.5 Stress notational convention

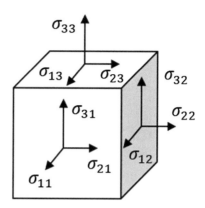

As an example, let us find traction $\mathbf{t}(\mathbf{n})$, normal stress vector $\mathbf{t}_n(\mathbf{n})$, and tangent stress vector $\mathbf{t}_t(\mathbf{n})$ for the given stress tensor [MPa]

$$\sigma = 7\mathbf{e}_1 \otimes \mathbf{e}_1 - 2(\mathbf{e}_1 \otimes \mathbf{e}_3 + \mathbf{e}_3 \otimes \mathbf{e}_1) + 5\mathbf{e}_2 \otimes \mathbf{e}_2 + 4\mathbf{e}_3 \otimes \mathbf{e}_3,$$

and area with normal

$$\mathbf{n} = \frac{2}{3}\mathbf{e}_1 - \frac{2}{3}\mathbf{e}_2 + \frac{1}{3}\mathbf{e}_3.$$

By the direct calculation we obtain

$$\mathbf{t} = \sigma\mathbf{n} = 7\mathbf{e}_1(\mathbf{e}_1 \cdot \mathbf{n}) - 2\mathbf{e}_1(\mathbf{e}_3 \cdot \mathbf{n}) - 2\mathbf{e}_3(\mathbf{e}_1 \cdot \mathbf{n}) + 5\mathbf{e}_2(\mathbf{e}_2 \cdot \mathbf{n}) + 4\mathbf{e}_3(\mathbf{e}_3 \cdot \mathbf{n}) = 4\mathbf{e}_1 - \frac{10}{3}\mathbf{e}_2,$$

$$\mathbf{t}_n = (\mathbf{t} \cdot \mathbf{n})\mathbf{n} = \left(4\mathbf{e}_1 \cdot \mathbf{n} - \frac{10}{3}\mathbf{e}_2 \cdot \mathbf{n}\right)\mathbf{n} = \frac{44}{9}\mathbf{n} = \frac{44}{27}(2\mathbf{e}_1 - 2\mathbf{e}_2 + \mathbf{e}_3),$$

$$\mathbf{t}_t = \mathbf{t} - \mathbf{t}_n = \frac{2}{27}(10\mathbf{e}_1 - \mathbf{e}_2 + 22\mathbf{e}_3).$$

3.3.3 Cauchy–Euler Equations

We return to the linear momentum balance (3.9), which can be rewritten using the stress tensor

$$\int \mathbf{b}_g dV = \oint \sigma\mathbf{n}\, dA. \tag{3.23}$$

Now, the divergence theorem (*Gauss–Green–Ostrogradskii*) allows us to transform the surface integral into the volumetric one

$$\oint \sigma \mathbf{n} dA = \int \mathrm{div}\sigma dV. \tag{3.24}$$

Substitution of (3.24) in (3.23) yields

$$\int (\mathbf{b}_g - \mathrm{div}\sigma) dV = \mathbf{0}. \tag{3.25}$$

Localizing it and using (3.10) we have, finally, the differential form of the linear momentum balance—*Cauchy–Euler equations*

$$\frac{d(\rho\mathbf{v})}{dt} + \rho\mathbf{v}\mathrm{div}\mathbf{v} = \mathrm{div}\sigma + \mathbf{b},$$

$$\frac{d(\rho v_i)}{dt} + \rho v_i \frac{\partial v_j}{\partial y_j} = \frac{\partial \sigma_{ij}}{\partial y_j} + b_i. \tag{3.26}$$

From the mathematical standpoint these are *partial differential equations* (PDEs) in contrast to the *ordinary differential equations* (ODEs) of the Newton law for volumeless mass-points. Thus, both initial and boundary conditions should be added to (3.26).

The *natural boundary conditions* represent the linear momentum balance on the surface $\partial\Omega$ of the body Ω and they can be written as follows

$$\sigma\mathbf{n} = \bar{\mathbf{t}}, \tag{3.27}$$

where the barred external traction $\bar{\mathbf{t}}$ is prescribed.

Alternatively, the *essential boundary conditions* imposed on placements $\bar{\mathbf{y}}$ can be prescribed on the boundary $\partial\Omega$

$$\mathbf{y} = \bar{\mathbf{y}}. \tag{3.28}$$

The *initial conditions* are
$$\mathbf{y}(t = 0) = \mathbf{y}_0,$$
$$\mathbf{v}(t = 0) = \mathbf{v}_0, \tag{3.29}$$

where $\mathbf{y}_0$ and $\mathbf{v}_0$ are prescribed in Ω.

3.4 Balance of Angular Momentum

As in the previous section we start with the mass-particle. Let us assume that $\mathbf{r}$ is a position vector of a mass-particle with respect to an arbitrarily chosen point. The cross product of the momentum balance equation with $\mathbf{r}$ gives

$$\mathbf{r} \times \frac{d}{dt}(m\mathbf{v}) = \mathbf{r} \times \mathbf{f}, \tag{3.30}$$

or

$$\frac{d}{dt}(\mathbf{r} \times m\mathbf{v}) = \mathbf{r} \times \mathbf{f}. \tag{3.31}$$

These two forms are equivalent because

$$\frac{d}{dt}(\mathbf{r} \times m\mathbf{v}) = \frac{d\mathbf{r}}{dt} \times (m\mathbf{v}) + \mathbf{r} \times \frac{d(m\mathbf{v})}{dt} = m\mathbf{v} \times \mathbf{v} + \mathbf{r} \times \frac{d(m\mathbf{v})}{dt} = \mathbf{r} \times \frac{d(m\mathbf{v})}{dt}.$$

In the case of continuum we have instead of (3.31)

$$\frac{d}{dt}\int \rho \mathbf{r} \times \mathbf{v} dV = \int \mathbf{r} \times \mathbf{b} dV + \oint \mathbf{r} \times \mathbf{t} dA. \tag{3.32}$$

It should not be missed by the reader that this equation of the *angular momentum* balance *cannot be derived* from the Euler law (3.7) as it was done for the Newton law in the case of mass-point. The angular momentum balance is an independent balance law for continua.

For the sake of further manipulations we rewrite the angular momentum balance in Cartesian coordinates

$$\varepsilon_{ijk}\left(\frac{d}{dt}\int \rho r_j v_k dV - \int r_j b_k dV - \oint r_j t_k dA\right) = 0. \tag{3.33}$$

The first and the third terms in the parentheses can be calculated by using the material time derivative of the volume integral and the divergence theorem accordingly

$$\frac{d}{dt}\int \rho r_j v_k dV = \int \left(\frac{d(\rho r_j v_k)}{dt} + \rho r_j v_k \frac{\partial v_m}{\partial y_m}\right) dV$$

$$= \int \left(r_j \frac{d(\rho v_k)}{dt} + \rho v_j v_k + \rho r_j v_k \frac{\partial v_m}{\partial y_m}\right) dV, \tag{3.34}$$

and

$$\oint r_j t_k dA = \oint r_j \sigma_{kl} n_l dA$$

$$= \int \frac{\partial(r_j \sigma_{kl})}{\partial y_l} dV$$

$$= \int \left(\delta_{jl}\sigma_{kl} + \frac{\partial \sigma_{kl}}{\partial y_l} r_j\right) dV$$

$$= \int \left(\sigma_{kj} + \frac{\partial \sigma_{kl}}{\partial y_l} r_j\right) dV. \tag{3.35}$$

Substituting these equations in the angular momentum balance we get

$$\varepsilon_{ijk}\int \left(r_j\left[\frac{d(\rho v_k)}{dt} + \rho v_k\frac{\partial v_m}{\partial y_m} - \frac{\partial \sigma_{kl}}{\partial y_l} - b_k\right] + \rho v_k v_j - \sigma_{kj}\right)dV = 0, \quad (3.36)$$

where the term in the brackets equals zero by virtue of the linear momentum balance. Thus, we remain with

$$\int \varepsilon_{ijk}(\rho v_k v_j - \sigma_{kj})dV = -\int \varepsilon_{ijk}\sigma_{kj}dV = 0. \qquad (3.37)$$

The latter equation can be obeyed for the symmetric Cauchy tensor only

$$\sigma_{kj} = \sigma_{jk}, \quad \boldsymbol{\sigma} = \boldsymbol{\sigma}^\mathrm{T}. \qquad (3.38)$$

The symmetry of the Cauchy stress tensor is the consequence of the angular momentum balance. This symmetry is not a law of Nature, however, and the Cauchy stress tensor can loose its symmetry in the case of electromechanical coupling, for example.

3.5 Master Balance Principle

All balance laws enjoy the same mathematical structure

$$\frac{d}{dt}\int \alpha dV = \int \xi dV + \oint \varphi \mathbf{n} dA, \qquad (3.39)$$

where ξ is the *volumetric supply* of α and φ is the *surface flux* of α.

Differentiating the integral and using the divergence theorem we localize the balance law

$$\frac{d\alpha}{dt} + \alpha \mathrm{div}\mathbf{v} = \mathrm{div}\varphi + \xi. \qquad (3.40)$$

The above considered three balance laws are summarized in Table 3.1.

Table 3.1 Balance laws in Eulerian description

	α	ξ	φ
Mass	ρ	–	–
Linear momentum	$\rho\mathbf{v}$	$\mathbf{b}$	σ
Angular momentum	$\rho\mathbf{r}\times\mathbf{v}$	$\mathbf{r}\times\mathbf{b}$	$\mathbf{r}\times\sigma$

3.6 Lagrangean Description

The description of balance laws was Eulerian because **y** was chosen as an independent variable. In the case of solids it is usually (but not always) more convenient to consider **x** as an independent variable, i.e. to use the Lagrangean description. The transition from one description to another is simple when the formulas relating material volumes and surfaces before and after deformation are used (see Sect. 2.1)

$$
\begin{aligned}
dV &= J dV_0, \\
d A \mathbf{n} &= J d A_0 \mathbf{F}^{-T} \mathbf{n}_0.
\end{aligned}
\tag{3.41}
$$

Substituting these equations in the master balance law (3.39) we get

$$
\frac{d}{dt} \int \alpha_0 dV_0 = \int \xi_0 dV_0 + \oint \varphi_0 \mathbf{n}_0 d A_0,
\tag{3.42}
$$

where we defined the Lagrangean quantities

$$
\begin{aligned}
\alpha_0(\mathbf{x}, t) &= J \alpha(\mathbf{y}(\mathbf{x}, t), t), \\
\xi_0(\mathbf{x}, t) &= J \xi(\mathbf{y}(\mathbf{x}, t), t), \\
\varphi_0(\mathbf{x}, t) &= J \varphi(\mathbf{y}(\mathbf{x}, t), t) \mathbf{F}^{-T}.
\end{aligned}
\tag{3.43}
$$

Localizing (3.42) we obtain

$$
\frac{\partial \alpha_0}{\partial t} = \mathrm{Div} \varphi_0 + \xi_0,
\tag{3.44}
$$

where "Div" operator is with respect to the referential coordinates **x**.

Particularly, the Lagrangean form of balance laws is summarized in Table 3.2, where

$$
\mathbf{P}(\mathbf{x}, t) = J \sigma(\mathbf{y}(\mathbf{x}, t), t) \mathbf{F}^{-T}
\tag{3.45}
$$

Table 3.2 Balance laws in Lagrangean description

	α_0	ξ_0	φ_0
Mass	ρ_0	–	–
Linear momentum	$\rho_0 \mathbf{v}$	$\mathbf{b}_0$	$\mathbf{P}$
Angular momentum	$\rho_0 \mathbf{r} \times \mathbf{v}$	$\mathbf{r} \times \mathbf{b}_0$	$\mathbf{r} \times \mathbf{P}$

is the *first Piola-Kirchhoff stress* tensor[1] (1PK);

$$\mathbf{b}_0 = J\mathbf{b} \tag{3.46}$$

is the body force per unit reference volume and

$$\rho_0 = J\rho \tag{3.47}$$

is the referential mass density.
 Mass balance takes form

$$\frac{\partial \rho_0}{\partial t} = 0. \tag{3.48}$$

Linear momentum balance is

$$\begin{aligned}\rho_0 \mathbf{a} &= \text{Div}\mathbf{P} + \mathbf{b}_0, \\ \rho_0 a_i &= \frac{\partial P_{ij}}{\partial x_j} + b_{0i},\end{aligned} \tag{3.49}$$

where $\mathbf{a} = \rho_0 \partial \mathbf{v}/\partial t$ is the acceleration vector.
 Angular momentum balance is

$$\mathbf{PF}^{\text{T}} = \mathbf{FP}^{\text{T}}. \tag{3.50}$$

Since the first Piola-Kirchhoff stress tensor is not symmetric it is convenient to introduce the *second Piola-Kirchhoff stress* tensor (2PK), which is symmetric,

$$\mathbf{S} = \mathbf{F}^{-1}\mathbf{P} = J\mathbf{F}^{-1}\sigma\mathbf{F}^{-\text{T}} = \mathbf{S}^{\text{T}}. \tag{3.51}$$

The natural boundary conditions in the Lagrangean setting read

$$\mathbf{Pn}_0 = \bar{\mathbf{t}}_0, \tag{3.52}$$

where $\mathbf{n}_0$ is the unit vector outward normal to the body surface $\partial \Omega_0$ in the referential configuration and $\bar{\mathbf{t}}_0$ is the prescribed Lagrangean traction on this surface. The Eulerian and Lagrangean quantities are related as follows

$$\begin{aligned}\mathbf{n} &= \mathbf{F}^{-\text{T}}\mathbf{n}_0 \left|\mathbf{F}^{-\text{T}}\mathbf{n}_0\right|^{-1}, \\ \mathbf{t} &= J^{-1}\mathbf{t}_0 \left|\mathbf{F}^{-\text{T}}\mathbf{n}_0\right|^{-1}.\end{aligned} \tag{3.53}$$

---

[1]Its transpose is often called the *nominal stress* tensor in the literature.

3.7 Lagrangean Equilibrium Equations in Curvilinear Coordinates

Lagrangean equilibrium equations in cylindrical and spherical coordinates for materials undergoing large deformations are rarely discussed in the literature. Nevertheless, these equations are very useful for solving nonlinear problems analytically or semi-analytically. So, our purpose in this section is to write down

$$\text{Div}\mathbf{P} = \mathbf{0} \tag{3.54}$$

in cylindrical and spherical coordinates.

We start with the divergence operator in cylindrical coordinates

$$\text{Div}\mathbf{P} = \frac{\partial \mathbf{P}}{\partial R}\mathbf{G}_R + \frac{1}{R}\frac{\partial \mathbf{P}}{\partial \Phi}\mathbf{G}_\Phi + \frac{\partial \mathbf{P}}{\partial Z}\mathbf{G}_Z = \mathbf{0}, \tag{3.55}$$

and calculate the divergence terms one by one for the first Piola-Kirchhoff stress

$$\begin{aligned}
\mathbf{P} &= P_{rR}\mathbf{g}_r \otimes \mathbf{G}_R + P_{r\phi}\mathbf{g}_r \otimes \mathbf{G}_\Phi + P_{rZ}\mathbf{g}_r \otimes GG_Z \\
&+ P_{\varphi R}\mathbf{g}_\varphi \otimes \mathbf{G}_R + P_{\varphi\phi}\mathbf{g}_\varphi \otimes \mathbf{G}_\Phi + P_{\varphi Z}\mathbf{g}_\varphi \otimes \mathbf{G}_Z \\
&+ P_{zR}\mathbf{g}_z \otimes \mathbf{G}_R + P_{z\phi}\mathbf{g}_z \otimes \mathbf{G}_\Phi + P_{zZ}\mathbf{g}_z \otimes \mathbf{G}_Z.
\end{aligned} \tag{3.56}$$

The first term of the divergence operator is calculated as follows

$$\begin{aligned}
\frac{\partial \mathbf{P}}{\partial R}\mathbf{G}_R &= \left(\frac{\partial P_{rR}}{\partial R}\mathbf{g}_r \otimes \mathbf{G}_R + P_{rR}\frac{\partial \mathbf{g}_r}{\partial R} \otimes \mathbf{G}_R + P_{rR}\mathbf{g}_r \otimes \frac{\partial \mathbf{G}_R}{\partial R} \right)\mathbf{G}_R \\
&+ \left(\frac{\partial P_{r\phi}}{\partial R}\mathbf{g}_r \otimes \mathbf{G}_\Phi + P_{r\phi}\frac{\partial \mathbf{g}_r}{\partial R} \otimes \mathbf{G}_\Phi + P_{r\phi}\mathbf{g}_r \otimes \frac{\partial \mathbf{G}_\Phi}{\partial R} \right)\mathbf{G}_R \\
&+ \left(\frac{\partial P_{rZ}}{\partial R}\mathbf{g}_r \otimes \mathbf{G}_Z + P_{rZ}\frac{\partial \mathbf{g}_r}{\partial R} \otimes \mathbf{G}_Z + P_{rZ}\mathbf{g}_r \otimes \frac{\partial \mathbf{G}_Z}{\partial R} \right)\mathbf{G}_R \\
&+ \left(\frac{\partial P_{\varphi R}}{\partial R}\mathbf{g}_\varphi \otimes \mathbf{G}_R + P_{\varphi R}\frac{\partial \mathbf{g}_\varphi}{\partial R} \otimes \mathbf{G}_R + P_{\varphi R}\mathbf{g}_\varphi \otimes \frac{\partial \mathbf{G}_R}{\partial R} \right)\mathbf{G}_R \\
&+ \left(\frac{\partial P_{\varphi\phi}}{\partial R}\mathbf{g}_\varphi \otimes \mathbf{G}_\Phi + P_{\varphi\phi}\frac{\partial \mathbf{g}_\varphi}{\partial R} \otimes \mathbf{G}_\Phi + P_{\varphi\phi}\mathbf{g}_\varphi \otimes \frac{\partial \mathbf{G}_\Phi}{\partial R} \right)\mathbf{G}_R \\
&+ \left(\frac{\partial P_{\varphi Z}}{\partial R}\mathbf{g}_\varphi \otimes \mathbf{G}_Z + P_{\varphi Z}\frac{\partial \mathbf{g}_\varphi}{\partial R} \otimes \mathbf{G}_Z + P_{\varphi Z}\mathbf{g}_\varphi \otimes \frac{\partial \mathbf{G}_Z}{\partial R} \right)\mathbf{G}_R \\
&+ \left(\frac{\partial P_{zR}}{\partial R}\mathbf{g}_z \otimes \mathbf{G}_R + P_{zR}\frac{\partial \mathbf{g}_z}{\partial R} \otimes \mathbf{G}_R + P_{zR}\mathbf{g}_z \otimes \frac{\partial \mathbf{G}_R}{\partial R} \right)\mathbf{G}_R \\
&+ \left(\frac{\partial P_{z\phi}}{\partial R}\mathbf{g}_z \otimes \mathbf{G}_\Phi + P_{z\phi}\frac{\partial \mathbf{g}_z}{\partial R} \otimes \mathbf{G}_\Phi + P_{z\phi}\mathbf{g}_z \otimes \frac{\partial \mathbf{G}_\Phi}{\partial R} \right)\mathbf{G}_R \\
&+ \left(\frac{\partial P_{zZ}}{\partial R}\mathbf{g}_z \otimes \mathbf{G}_Z + P_{zZ}\frac{\partial \mathbf{g}_z}{\partial R} \otimes \mathbf{G}_Z + P_{zZ}\mathbf{g}_z \otimes \frac{\partial \mathbf{G}_Z}{\partial R} \right)\mathbf{G}_R.
\end{aligned} \tag{3.57}$$

This equation reduces to

$$\frac{\partial \mathbf{P}}{\partial R}\mathbf{G}_R = \frac{\partial P_{rR}}{\partial R}\mathbf{g}_r + P_{rR}\frac{\partial \mathbf{g}_r}{\partial R} + \frac{\partial P_{\varphi R}}{\partial R}\mathbf{g}_\varphi$$
$$+ P_{\varphi R}\frac{\partial \mathbf{g}_\varphi}{\partial R} + \frac{\partial P_{zR}}{\partial R}\mathbf{g}_z + P_{zR}\frac{\partial \mathbf{g}_z}{\partial R}, \tag{3.58}$$

with account of derivatives

$$\frac{\partial \mathbf{G}_R}{\partial R} = \frac{\partial \mathbf{G}_\Phi}{\partial \Phi} = \frac{\partial \mathbf{G}_Z}{\partial R} = \frac{\partial \mathbf{G}_Z}{\partial \Phi} = \frac{\partial \mathbf{G}_R}{\partial Z} = \frac{\partial \mathbf{G}_\Phi}{\partial Z} = \frac{\partial \mathbf{G}_Z}{\partial Z} = \mathbf{0},$$
$$\frac{\partial \mathbf{G}_R}{\partial \Phi} = \mathbf{G}_\Phi, \quad \frac{\partial \mathbf{G}_\Phi}{\partial \Phi} = -\mathbf{G}_R. \tag{3.59}$$

Derivatives of the Eulerian basis vectors with respect to the Lagrangean coordinates are given in (2.20). By using them we finally get

$$\frac{\partial \mathbf{P}}{\partial R}\mathbf{G}_R = \left(\frac{\partial P_{rR}}{\partial R} - P_{\varphi R}\frac{\partial \varphi}{\partial R} \right) \mathbf{g}_r$$
$$+ \left(P_{rR}\frac{\partial \varphi}{\partial R} + \frac{\partial P_{\varphi R}}{\partial R} \right) \mathbf{g}_\varphi$$
$$+ \frac{\partial P_{zR}}{\partial R}\mathbf{g}_z. \tag{3.60}$$

In a similar way we calculate

$$\frac{1}{R}\frac{\partial \mathbf{P}}{\partial \Phi}\mathbf{G}_\Phi = \left(\frac{P_{rR}}{R} + \frac{1}{R}\frac{\partial P_{r\Phi}}{\partial \Phi} - \frac{P_{\varphi\Phi}}{R}\frac{\partial \varphi}{\partial \Phi} \right) \mathbf{g}_r$$
$$+ \left(\frac{P_{r\Phi}}{R}\frac{\partial \varphi}{\partial \Phi} + \frac{P_{\varphi R}}{R} + \frac{1}{R}\frac{\partial P_{\varphi\Phi}}{\partial \Phi} \right) \mathbf{g}_\varphi$$
$$+ \left(\frac{P_{zR}}{R} + \frac{1}{R}\frac{\partial P_{z\Phi}}{\partial \Phi} \right) \mathbf{g}_z, \tag{3.61}$$

and

$$\frac{\partial \mathbf{P}}{\partial Z}\mathbf{G}_Z = \left(\frac{\partial P_{rZ}}{\partial Z} - P_{\varphi Z}\frac{\partial \varphi}{\partial Z} \right) \mathbf{g}_r$$
$$+ \left(\frac{\partial P_{\varphi Z}}{\partial Z} + P_{rZ}\frac{\partial \varphi}{\partial Z} \right) \mathbf{g}_\varphi$$
$$+ \frac{\partial P_{zZ}}{\partial Z}\mathbf{g}_z. \tag{3.62}$$

Substitution of (3.60), (3.61), (3.62) in (3.54) yields a system of three scalar Lagrangean equilibrium equations in cylindrical coordinates

$$\frac{\partial P_{rR}}{\partial R} - P_{\varphi R}\frac{\partial \varphi}{\partial R} + \frac{P_{rR}}{R} + \frac{1}{R}\frac{\partial P_{r\Phi}}{\partial \Phi} - \frac{P_{\varphi \Phi}}{R}\frac{\partial \varphi}{\partial \Phi} + \frac{\partial P_{rZ}}{\partial Z} - P_{\varphi Z}\frac{\partial \varphi}{\partial Z} = 0,$$

$$P_{rR}\frac{\partial \varphi}{\partial R} + \frac{\partial P_{\varphi R}}{\partial R} + \frac{P_{r\Phi}}{R}\frac{\partial \varphi}{\partial \Phi} + \frac{P_{\varphi R}}{R} + \frac{1}{R}\frac{\partial P_{\varphi \Phi}}{\partial \Phi} + \frac{\partial P_{\varphi Z}}{\partial Z} + P_{rZ}\frac{\partial \varphi}{\partial Z} = 0, \quad (3.63)$$

$$\frac{\partial P_{zR}}{\partial R} + \frac{P_{zR}}{R} + \frac{1}{R}\frac{\partial P_{z\Phi}}{\partial \Phi} + \frac{\partial P_{zZ}}{\partial Z} = 0.$$

Analogously, it is possible to obtain (left for the exercise) the Lagrangean equilibrium equations in spherical coordinates

$$\frac{\partial P_{rR}}{\partial R} - P_{\theta R}\frac{\partial \theta}{\partial R} - sP_{\varphi R}\frac{\partial \varphi}{\partial R} + 2\frac{P_{rR}}{R} + \frac{1}{R}\frac{\partial P_{r\Theta}}{\partial \Theta} - \frac{P_{\theta\Theta}}{R}\frac{\partial \theta}{\partial \Theta}$$
$$- s\frac{P_{\varphi\Theta}}{R}\frac{\partial \varphi}{\partial \Theta} + \frac{CP_{r\Theta}}{RS} + \frac{1}{RS}\frac{\partial P_{r\Phi}}{\partial \Phi} - \frac{P_{\theta\Phi}}{RS}\frac{\partial \theta}{\partial \Phi} - \frac{sP_{\varphi\Phi}}{RS}\frac{\partial \varphi}{\partial \Phi} = 0,$$

$$\frac{\partial P_{\theta R}}{\partial R} + P_{rR}\frac{\partial \theta}{\partial R} - cP_{\varphi R}\frac{\partial \varphi}{\partial R} + 2\frac{P_{\theta R}}{R} + \frac{1}{R}\frac{\partial P_{\theta\Theta}}{\partial \Theta} + \frac{P_{r\Theta}}{R}\frac{\partial \theta}{\partial \Theta}$$
$$- c\frac{P_{\varphi\Theta}}{R}\frac{\partial \varphi}{\partial \Theta} + \frac{CP_{\theta\Theta}}{RS} + \frac{1}{RS}\frac{\partial P_{\theta\Phi}}{\partial \Phi} + \frac{P_{r\Phi}}{RS}\frac{\partial \theta}{\partial \Phi} - \frac{cP_{\varphi\Phi}}{RS}\frac{\partial \varphi}{\partial \Phi} = 0, \quad (3.64)$$

$$\frac{\partial P_{\varphi R}}{\partial R} + sP_{rR}\frac{\partial \varphi}{\partial R} + cP_{\theta R}\frac{\partial \varphi}{\partial R} + 2\frac{P_{\varphi R}}{R} + \frac{1}{R}\frac{\partial P_{\varphi\Theta}}{\partial \Theta} + s\frac{P_{r\Theta}}{R}\frac{\partial \varphi}{\partial \Theta}$$
$$+ c\frac{P_{\theta\Theta}}{R}\frac{\partial \varphi}{\partial \Theta} + \frac{CP_{\varphi\Theta}}{RS} + \frac{1}{RS}\frac{\partial P_{\varphi\Phi}}{\partial \Phi} + \frac{sP_{r\Phi}}{RS}\frac{\partial \varphi}{\partial \Phi} + \frac{cP_{\theta\Phi}}{RS}\frac{\partial \varphi}{\partial \Phi} = 0,$$

where
$$S = \sin\Theta, \quad C = \cos\Theta, \quad s = \sin\theta, \quad c = \cos\theta.$$

3.8 Exercises

1. Derive (3.61).
2. Derive (3.62).
3. Derive (3.64).

References

Chadwick P (1999) Continuum mechanics. Dover, New York

Truesdell C, Toupin RA (1960) Classical field theories. In: Flugge S (ed) Encyclopedia of physics, vol III/1. Springer, New York

Volokh KY (2006) Lagrangean equilibrium equations in cylindrical and spherical coordinates. Comput Mater Continua 3:37–42

Chapter 4
Isotropic Elasticity

Balance laws are equally applicable to all materials. No specification of material has been done yet. Thus, physics requires the development of additional equations that characterize material behavior. The requirement of the additional—*constitutive equations*—also comes from mathematics because of the need to close the system of governing equations in which six equations are lacking—Table 4.1.

Thus, the purpose of this chapter is to introduce the constitutive theory of elasticity (hyperelasticity) for isotropic materials.

4.1 Strain Energy

The *rheological model* for elastic material is a spring—Fig. 4.1.

For the classical *linear* spring, stress σ is equal to (small) strain ε scaled by Young modulus E,

$$\sigma = E\varepsilon.$$

This equation is called Hooke's law in honor of Robert Hooke.

Evidently, this constitutive law is a linearization of a more general function describing a nonlinear spring

$$\sigma(\varepsilon).$$

Although this function can be fitted in experiments only, it is possible to draw some conclusions about it considering the work of stress on strain

$$\psi = \int \sigma(\varepsilon)d\varepsilon.$$

© Springer Science+Business Media Singapore 2016
K. Volokh, *Mechanics of Soft Materials*, DOI 10.1007/978-981-10-1599-1_4

Table 4.1 Equations versus variables

Equations	Variables
Mass conservation: 1	Mass density: 1
Linear Momentum: 3	Stress tensor: 9
Angular momentum: 3	Placement vector: 3
Sum: 7	Sum: 13

Fig. 4.1 Spring as a rheological model of elasticity

In the case of an ideal elastic spring, this work does not depend on the loading history and it only depends on the initial and final states of the spring—the integration limits. If the integral is path-independent then the integrand should be a full differential and, consequently,

$$\sigma = \frac{d\psi}{d\varepsilon}. \tag{4.1}$$

For example, in the case of the Hookean elasticity we have the *stored* or *strain energy function*

$$\psi = \frac{1}{2}E\varepsilon^2.$$

Elastic material is called *hyperelastic* if its constitutive equation can be defined by a strain energy function—(4.1). In this book, we do not make distinction between elastic and hyperelastic materials. Such distinction would be difficult to justify from the physics standpoint.

The extension of the simplistic formula (4.1) to three dimensions is not trivial. Indeed, variety of stresses and strains can be considered and it is not clear which stress works on which strain. To clarify that we consider the work of external forces on displacement increments, $d\mathbf{u} = d\mathbf{y}$, over volume V_0

$$d\Pi = \oint \bar{\mathbf{t}}_0 \cdot d\mathbf{y} dA_0 + \int (\mathbf{b}_0 - \rho_0 \mathbf{a}) \cdot d\mathbf{y} dV_0, \tag{4.2}$$

where $\bar{\mathbf{t}}_0$ and $(\mathbf{b}_0 - \rho_0 \mathbf{a})$ designate prescribed tractions per the reference area and generalized body forces per the reference volume respectively.

By using (3.49) we can rewrite the energy increment in the form

$$d\Pi = \oint \bar{\mathbf{t}}_0 \cdot d\mathbf{y} dA_0 - \int \text{Div}\mathbf{P} \cdot d\mathbf{y} dV_0$$

$$= \oint \bar{t}_{0i} dy_i dA_0 - \int \frac{\partial P_{ij}}{\partial x_j} dy_i dV_0$$

$$= \oint \bar{t}_{0i} dy_i dA_0 - \int \frac{\partial(P_{ij} dy_i)}{\partial x_j} dV_0 + \int P_{ij} \frac{\partial(dy_i)}{\partial x_j} dV_0$$

$$= \oint (\bar{t}_{0i} - P_{ij} n_{0j}) dy_i dA_0 + \int P_{ij} d\frac{\partial y_i}{\partial x_j} dV_0$$

$$= \oint (\bar{\mathbf{t}}_0 - \mathbf{P} \mathbf{n}_0) \cdot d\mathbf{y} dA_0 + \int \mathbf{P} : d\mathbf{F} V_0$$

$$= \int \mathbf{P} : d\mathbf{F} dV_0, \tag{4.3}$$

where the traction boundary conditions, $\mathbf{P}\mathbf{n}_0 = \bar{\mathbf{t}}_0$, is used.

This formula means that the incremental work of the external forces is equal to the incremental work of the internal forces. The work of the internal forces per unit reference volume can be designated as follows

$$d\psi = \mathbf{P} : d\mathbf{F}. \tag{4.4}$$

Analogously to (4.1) the work is path-independent only in the case of full differential $d\psi$ which implies

$$\mathbf{P} = \frac{\partial \psi}{\partial \mathbf{F}}, \quad P_{ij} = \frac{\partial \psi}{\partial F_{ij}}. \tag{4.5}$$

Here ψ is called the *stored* or *strain energy per unit reference volume* and material obeying (4.5) is called *hyperelastic*.

Evidently, the first Piola-Kirchhoff stress makes a *work-conjugate* couple with the deformation gradient. It is possible, however, to assume that the strain energy depends on the Green strain, $\mathbf{E} = \frac{1}{2}(\mathbf{F}^{\mathrm{T}}\mathbf{F} - \mathbf{1})$, rather than on the deformation gradient. In this case we have by using the chain rule for differentiation

$$\mathbf{P} = \frac{\partial \psi}{\partial \mathbf{E}} : \frac{\partial \mathbf{E}}{\partial \mathbf{F}} = \mathbf{F}\frac{\partial \psi}{\partial \mathbf{E}}. \tag{4.6}$$

On the other hand we have from (3.51)

$$\mathbf{P} = \mathbf{F}\mathbf{S}, \tag{4.7}$$

where $\mathbf{S}$ is the second Piola-Kirchhoff stress tensor and, consequently,

$$\mathbf{S} = \frac{\partial \psi}{\partial \mathbf{E}}, \tag{4.8}$$

or

$$\mathbf{S} = 2\frac{\partial \psi}{\partial \mathbf{C}}, \tag{4.9}$$

where $\mathbf{C} = \mathbf{F}^\mathrm{T}\mathbf{F} = 2\mathbf{E} + \mathbf{1}$ is the right Cauchy–Green tensor.

It is possible to show by a direct calculation that the considered stress-strain pairs are work-conjugate

$$\mathbf{P} : d\mathbf{F} = \mathbf{S} : d\mathbf{E}. \tag{4.10}$$

The Cauchy stress tensor is obtained from (3.45) as follows

$$\sigma = J^{-1}\mathbf{P}\mathbf{F}^\mathrm{T} = J^{-1}\mathbf{F}\mathbf{S}\mathbf{F}^\mathrm{T} = 2J^{-1}\mathbf{F}\frac{\partial\psi}{\partial\mathbf{C}}\mathbf{F}^\mathrm{T}. \tag{4.11}$$

We showed that the strain energy could be defined as a function of various strains. Is there any preference in the choice of strains? The answer is yes. The strains which are insensitive to the superimposed rigid body motion (RBM) are necessary to suppress stressing without deformation. The latter means that the strain energy should obey the following condition

$$\psi(\mathbf{F}) = \psi(\mathbf{Q}\mathbf{F}), \tag{4.12}$$

where $\mathbf{Q}$ is a rotation tensor.

Expectedly, the Lagrangean strain tensors are not affected by the superimposed RBM

$$\mathbf{C}^* = \mathbf{F}^{*\mathrm{T}}\mathbf{F}^* = (\mathbf{Q}\mathbf{F})^\mathrm{T}(\mathbf{Q}\mathbf{F}) = \mathbf{F}\mathbf{Q}^\mathrm{T}\mathbf{Q}\mathbf{F} = \mathbf{F}^\mathrm{T}\mathbf{F} = \mathbf{C},$$
$$\mathbf{E}^* = \frac{1}{2}(\mathbf{C}^* - \mathbf{1}) = \frac{1}{2}(\mathbf{C} - \mathbf{1}) = \mathbf{E}. \tag{4.13}$$

The Cauchy stress is objective (see Sect. 2.6)

$$\sigma(\mathbf{Q}\mathbf{F}) = 2J^{-1}\mathbf{Q}\mathbf{F}\frac{\partial\psi}{\partial\mathbf{C}}\mathbf{F}^\mathrm{T}\mathbf{Q}^\mathrm{T} = \mathbf{Q}\sigma(\mathbf{F})\mathbf{Q}^\mathrm{T}. \tag{4.14}$$

4.2   Strain Energy Depending on Invariants

Ronald Rivlin significantly contributed to the development of the following representation for the strain energy of isotropic materials (given without proof)

$$\psi(\mathbf{C}) = \psi(I_1, I_2, I_3), \tag{4.15}$$

$$I_1 = \mathrm{tr}\mathbf{C}, \quad I_2 = \frac{1}{2}\{(\mathrm{tr}\mathbf{C})^2 - \mathrm{tr}(\mathbf{C}^2)\}, \quad I_3 = \det\mathbf{C}. \tag{4.16}$$

That is the strain energy depends on the invariants of the right Cauchy–Green tensor. Based on this representation we can calculate the constitutive law as follows

$$\mathbf{S} = 2\frac{\partial \psi}{\partial \mathbf{C}} = 2\left(\frac{\partial \psi}{\partial I_1}\frac{\partial I_1}{\partial \mathbf{C}} + \frac{\partial \psi}{\partial I_2}\frac{\partial I_2}{\partial \mathbf{C}} + \frac{\partial \psi}{\partial I_3}\frac{\partial I_3}{\partial \mathbf{C}}\right). \tag{4.17}$$

The tensor derivatives of invariants have been obtained in Sect. 1.3 and we can write them down for symmetric tensor $\mathbf{C}$

$$\frac{\partial I_1}{\partial \mathbf{C}} = \mathbf{1}, \quad \frac{\partial I_2}{\partial \mathbf{C}} = I_1\mathbf{1} - \mathbf{C}, \quad \frac{\partial I_3}{\partial \mathbf{C}} = I_3\mathbf{C}^{-1}. \tag{4.18}$$

Thus, the constitutive law takes the canonical form

$$\mathbf{S} = 2\left\{\left(\frac{\partial \psi}{\partial I_1} + I_1\frac{\partial \psi}{\partial I_2}\right)\mathbf{1} - \frac{\partial \psi}{\partial I_2}\mathbf{C} + I_3\frac{\partial \psi}{\partial I_3}\mathbf{C}^{-1}\right\}, \tag{4.19}$$

or for the Cauchy stress

$$\boldsymbol{\sigma} = J^{-1}\mathbf{F}\mathbf{S}\mathbf{F}^{\mathrm{T}} = 2J^{-1}\left\{\left(\frac{\partial \psi}{\partial I_1} + I_1\frac{\partial \psi}{\partial I_2}\right)\mathbf{B} - \frac{\partial \psi}{\partial I_2}\mathbf{B}^2 + I_3\frac{\partial \psi}{\partial I_3}\mathbf{1}\right\}, \tag{4.20}$$

where $\mathbf{B} = \mathbf{F}\mathbf{F}^{\mathrm{T}}$ is the left Cauchy–Green tensor.

We remind (see Sect. 2.3) that invariants of $\mathbf{B}$ coincide with the invariants of $\mathbf{C}$: $I_a = I_a(\mathbf{C}) = I_a(\mathbf{B})$. The latter notion allows us to rewrite the constitutive equation for the Cauchy stress in the compact form

$$\boldsymbol{\sigma} = 2J^{-1}\frac{\partial \psi}{\partial \mathbf{B}}\mathbf{B}. \tag{4.21}$$

4.3 Strain Energy Depending on Principal Stretches

Alternatively to the use of invariants it is possible to formulate the strain energy in terms of principal stretches

$$\psi(\mathbf{C}) = \psi(\lambda_1, \lambda_2, \lambda_3). \tag{4.22}$$

Derivation of the constitutive law in principal stretches is a bit tricky. We start with the calculation of the energy increment as follows

$$d\psi(\lambda_1, \lambda_2, \lambda_3) = \frac{\partial \psi}{\partial \lambda_1}d\lambda_1 + \frac{\partial \psi}{\partial \lambda_2}d\lambda_2 + \frac{\partial \psi}{\partial \lambda_3}d\lambda_3. \tag{4.23}$$

In order to find $d\lambda_i$ we, firstly, calculate the increment of the spectral decomposition (2.25)

$$dC = \sum_{a=1}^{3} \left(2\lambda_a d\lambda_a \mathbf{m}^{(a)} \otimes \mathbf{m}^{(a)} + \lambda_a^2 d\mathbf{m}^{(a)} \otimes \mathbf{m}^{(a)} + \lambda_a^2 \mathbf{m}^{(a)} \otimes d\mathbf{m}^{(a)}\right). \quad (4.24)$$

Secondly, we project $d\mathbf{C}$ on $\mathbf{m}^{(1)} \otimes \mathbf{m}^{(1)}$ as follows

$$\mathbf{m}^{(1)} \otimes \mathbf{m}^{(1)} : d\mathbf{C} = 2\lambda_1 d\lambda_1, \quad (4.25)$$

where we used: $\mathbf{m}^{(1)} \cdot \mathbf{m}^{(a)} = \delta_{1a}$ and $d(\mathbf{m}^{(1)} \cdot \mathbf{m}^{(1)}) = 0 \Rightarrow d\mathbf{m}^{(1)} \cdot \mathbf{m}^{(1)} = 0$.
Thus, we have

$$d\lambda_1 = \frac{1}{2\lambda_1}\mathbf{m}^{(1)} \otimes \mathbf{m}^{(1)} : d\mathbf{C}. \quad (4.26)$$

Analogously, projecting $d\mathbf{C}$ on $\mathbf{m}^{(2)} \otimes \mathbf{m}^{(2)}$ and $\mathbf{m}^{(3)} \otimes \mathbf{m}^{(3)}$ we obtain

$$d\lambda_2 = \frac{1}{2\lambda_2}\mathbf{m}^{(2)} \otimes \mathbf{m}^{(2)} : d\mathbf{C},$$
$$d\lambda_3 = \frac{1}{2\lambda_3}\mathbf{m}^{(3)} \otimes \mathbf{m}^{(3)} : d\mathbf{C}. \quad (4.27)$$

Substituting all stretch increments into the energy increment we get

$$d\psi = \frac{\partial \psi}{\partial \mathbf{C}} : d\mathbf{C}, \quad (4.28)$$

where

$$\frac{\partial \psi}{\partial \mathbf{C}} = \sum_{a=1}^{3} \frac{1}{2\lambda_a} \frac{\partial \psi}{\partial \lambda_a} \mathbf{m}^{(a)} \otimes \mathbf{m}^{(a)}. \quad (4.29)$$

Using this derivative we can write the constitutive law for the second Piola-Kirchhoff stress tensor in the form

$$\mathbf{S} = 2\frac{\partial \psi}{\partial \mathbf{C}} = \sum_{a=1}^{3} \frac{1}{\lambda_a} \frac{\partial \psi}{\partial \lambda_a} \mathbf{m}^{(a)} \otimes \mathbf{m}^{(a)}. \quad (4.30)$$

It is remarkable that the second Piola-Kirchhoff stress is coaxial with the right Cauchy–Green tensor because their principal directions coincide. The latter allows the direct calculation of the principal second Piola-Kirchhoff stresses

$$S_a = \frac{1}{\lambda_a} \frac{\partial \psi}{\partial \lambda_a} \quad \text{(no sum)}. \quad (4.31)$$

By using the spectral decomposition of the deformation gradient, (2.39), we can compute the Cauchy stress as follows

$$\sigma = \frac{1}{\lambda_1 \lambda_2 \lambda_3} \sum_{a=1}^{3} \lambda_a \frac{\partial \psi}{\partial \lambda_a} \mathbf{n}^{(a)} \otimes \mathbf{n}^{(a)}, \tag{4.32}$$

which is coaxial with the left Cauchy–Green tensor because their principal directions (eigenvectors) coincide. The latter allows the direct computation of the principal Cauchy stresses

$$\sigma_a = \frac{\lambda_a}{\lambda_1 \lambda_2 \lambda_3} \frac{\partial \psi}{\partial \lambda_a} \quad \text{(no sum)}. \tag{4.33}$$

4.4 Incompressibility

Many soft materials resist volume changes much stronger than the shape changes. This experimental observation makes it reasonable to assume the material *incompressibility*

$$\frac{dV}{dV_0} = J = \det \mathbf{F} = 1 = \det \mathbf{B} = \det \mathbf{C} = I_3. \tag{4.34}$$

This can be considered as a restriction imposed on deformation, which is called *isochoric* in this case,

$$\gamma(\mathbf{C}) = 1 - I_3(\mathbf{C}) = 0. \tag{4.35}$$

The increment of the restriction is

$$d\gamma(\mathbf{C}) = \frac{\partial \gamma}{\partial \mathbf{C}} : d\mathbf{C} = 0. \tag{4.36}$$

Here $\partial \gamma / \partial \mathbf{C}$ can be interpreted as a stress producing zero work on the strain increment the *workless stress*. Such stress is indefinite since it can always be scaled by an arbitrary parameter Π.

Adding the workless stress to the stress derived from the strain energy we have the constitutive law in the form

$$\sigma = 2J^{-1} \mathbf{F} \left(\frac{\partial \psi}{\partial \mathbf{C}} + \Pi \frac{\partial \gamma}{\partial \mathbf{C}} \right) \mathbf{F}^{\mathrm{T}}, \tag{4.37}$$

or, substituting for γ from (4.35)

$$\sigma = 2J^{-1} \mathbf{F} \frac{\partial \psi}{\partial \mathbf{C}} \mathbf{F}^{\mathrm{T}} - \Pi \mathbf{1}. \tag{4.38}$$

The unknown parameter Π should be obtained from the solution of balance equations.

In the case of isotropic material we have

$$\sigma = -\Pi\mathbf{1} + 2(\psi_1 + I_1\psi_2)\mathbf{B} - 2\psi_2\mathbf{B}^2, \tag{4.39}$$

where

$$\psi_a \equiv \frac{\partial\psi}{\partial I_a}. \tag{4.40}$$

In terms of the principal stresses and stretches we have

$$\sigma_a = \lambda_a \frac{\partial\psi}{\partial\lambda_a} - \Pi \quad \text{(no sum)}. \tag{4.41}$$

Finally, we should note that the strict account of the material incompressibility presented in this section is usually a blessing for analytical and semi-analytical solutions. The latter approach was mastered by Ronald Samuel Rivlin (Barenblatt and Joseph 1997). At the same time the incompressibility constraint can be a pain in the neck for numerical methods. In the latter case various penalty approaches are preferable as compared to the formulation given above.

4.5   Examples of Strain Energy

In this section we consider some popular strain energy functions $\psi(\mathbf{C})$, which in the absence of *residual stresses* should meet the following conditions for energy

$$\psi(\mathbf{1}) = 0, \quad \frac{\partial\psi}{\partial\mathbf{C}}(\mathbf{1}) = \mathbf{0}, \tag{4.42}$$

or, in the case of incompressible material

$$\frac{\partial\psi}{\partial\mathbf{C}}(\mathbf{1}) - \Pi_0\mathbf{1} = \mathbf{0}, \tag{4.43}$$

where Π_0 is the specialization of Π in the referential state.

We start with the *Kirchhoff–Saint Venant* material

$$\psi(\mathbf{E}) = \frac{c_1}{2}(\text{tr}\mathbf{E})^2 + c_2\mathbf{E} : \mathbf{E}, \tag{4.44}$$

where c_1 and c_2 are the material (Lame) constants and the Green strain is $\mathbf{E} = \frac{1}{2}(\mathbf{C} - \mathbf{1})$. Differentiating the strain energy density with respect to the Green strain we obtain 2PK stress

$$\begin{aligned}
S_{ij} = \frac{\partial \psi}{\partial E_{ij}} &= \frac{c_1}{2}\frac{\partial(E_{kk}E_{rr})}{\partial E_{ij}} + c_2\frac{\partial(E_{mn}E_{mn})}{\partial E_{ij}} \\
&= \frac{c_1}{2}\frac{\partial E_{kk}}{\partial E_{ij}}E_{rr} + 2c_2\frac{\partial E_{mn}}{\partial E_{ij}}E_{mn} \\
&= \frac{c_1}{2}\delta_{ki}\delta_{kj}E_{rr} + 2c_2\delta_{mi}\delta_{nj}E_{mn} \\
&= \frac{c_1}{2}\delta_{ij}E_{rr} + 2c_2 E_{ij},
\end{aligned}$$

or

$$\mathbf{S} = \frac{\partial \psi}{\partial \mathbf{E}} = \frac{c_1}{2}(\mathrm{tr}\mathbf{E})\mathbf{1} + 2c_2\mathbf{E}. \tag{4.45}$$

Alternatively, we can express the strain energy in terms of principal stretches

$$\psi(\lambda_1, \lambda_2, \lambda_3) = \frac{c_1}{8}(\lambda_1^2+\lambda_2^2+\lambda_3^2-3)^2 + \frac{c_2}{4}\{(\lambda_1^2-1)^2+(\lambda_2^2-1)^2+(\lambda_3^2-1)^2\}, \tag{4.46}$$

and constitutive equations in terms of principal stresses and stretches are

$$S_a = \frac{c_1}{2}(\lambda_1^2 + \lambda_2^2 + \lambda_3^2 - 3) + c_2(\lambda_a^2 - 1), \quad a = 1, 2, 3. \tag{4.47}$$

The Kirchhoff–Saint Venant material model is obtained from the classical Hooke model by replacing small strain tensor with the Green strain tensor. This model is rarely used for soft materials per se. Nevertheless, it might be useful for hard materials undergoing small strains in order to suppress the effect of the rigid body motion in numerical computations. Indeed, in the case of the rigid body motion we have: $\mathbf{y} = \mathbf{Q}(t)\mathbf{x}+\mathbf{c}(t)$, where $\mathbf{Q}^{\mathrm{T}} = \mathbf{Q}^{-1}$ (det $\mathbf{Q} = 1$) is the rotation tensor and $\mathbf{c}$ is a vector of translation. Its deformation gradient is $\mathbf{F} = \mathrm{Grad}\mathbf{y} = \mathbf{Q}$, and the displacement gradient is $\mathbf{H} = \mathrm{Grad}\mathbf{u} = \mathrm{Grad}(\mathbf{y} - \mathbf{x}) = \mathbf{F} - \mathbf{1} = \mathbf{Q} - \mathbf{1}$. Thus, the tensor of small deformations is not zero: $\frac{1}{2}(\mathbf{H} + \mathbf{H}^{\mathrm{T}}) = \frac{1}{2}(\mathbf{Q}+\mathbf{Q}^{\mathrm{T}} - 2\mathbf{1}) \neq \mathbf{0}$; while the finite deformation Green strain is zero: $\mathbf{E} = \frac{1}{2}(\mathbf{F}^{\mathrm{T}}\mathbf{F} - \mathbf{1}) = \frac{1}{2}(\mathbf{Q}^{\mathrm{T}}\mathbf{Q} - \mathbf{1}) = \frac{1}{2}(\mathbf{1} - \mathbf{1}) = \mathbf{0}$.

Next strain energy function defines the *neo-Hookean* incompressible material

$$\psi = c_1(I_1 - 3), \quad J = 1, \tag{4.48}$$

where c_1 is a material constant.

The neo-Hookean model is the simplest one for soft materials. It is often used as a starting point for the experimental calibration. Actually, this model works well for moderate stretches up to ~ 1.5 for rubberlike materials.

A popular generalization of the neo-Hookean model is the *Yeoh* material defined as a polynomial of the first principal invariant I_1

$$\psi = c_1(I_1 - 3) + c_2(I_1 - 3)^2 + c_3(I_1 - 3)^3, \quad J = 1, \tag{4.49}$$

where the material constants are calibrated by Hamdi et al. (2006), for example, for natural rubber (NR)

$$c_1 = 0.298 \text{ MPa}, \quad c_2 = 0.014 \text{ MPa}, \quad c_3 = 0.00016 \text{ MPa}.$$

The Yeoh model can describe large stretches of rubberlike materials up to the point of failure.

Another generalization of the neo-Hookean model is the *Mooney–Rivlin* material which defines the dependence of the strain energy on both the first and second principal invariants

$$\psi = c_1(I_1 - 3) + c_2(I_2 - 3) + c_3(I_1 - 3)^2 + c_4(I_1 - 3)(I_2 - 3) + c_5(I_2 - 3)^2, \quad J = 1, \tag{4.50}$$

where the material constants are calibrated by Sasso et al. (2008), for example,

$$c_1 = 0.59 \text{ MPa}, \ c_2 = -0.039 \text{ MPa}, \ c_3 = -0.0028 \text{ MPa}, \ c_4 = 0.0076 \text{ MPa}, \ c_5 = -0.00077 \text{ MPa}.$$

Further generalization of the previous models is the *Ogden* material defined in terms of principal stretches

$$\psi = \sum_{p=1}^{N} \frac{\mu_p}{\alpha_p} \left(\lambda_1^{\alpha_p} + \lambda_2^{\alpha_p} + \lambda_3^{\alpha_p} - 3 \right), \quad J = 1, \quad \mu_p \alpha_p > 0, \quad p = 1, \ldots N, \tag{4.51}$$

where the material constants are calibrated by Hamdi et al. (2006), for example, for styrene-butadiene rubber (SBR)

$$N = 2, \ \mu_1 = 0.638 \text{ MPa}, \ \alpha_1 = 3.03, \ \mu_2 = -0.025 \text{ MPa}, \ \alpha_2 = -2.35.$$

We considered very few material models to give a taste of them. Generally, the number of possible constitutive models is approximately equal to the number of researchers who work (or worked) in the field.

4.6 Energy Limiter

It is worth noting that the classical models of hyperelasticity are devoted to the *intact* material behavior, in which the strain energy goes to infinity $\psi \to \infty$ with the increasing strain $\|\mathbf{F}\| \to \infty$ (where $\|...\|$ is a tensor norm). However, no real material can accumulate energy infinitely. The strain energy must be bounded. In this section, we introduce an *energy limiter* in the expression for strain energy. Such limiter enforces saturation—the failure energy—in the strain energy function, which indicates the maximum amount of energy that can be stored and dissipated by an infinitesimal material volume. The limiter induces *stress bounds* in the constitutive equations *automatically*.

We use the following form of the strain energy function

$$\psi(\mathbf{F}, \alpha) = \psi_f - H(\alpha)\psi_e(\mathbf{F}), \tag{4.52}$$

where

$$\psi_f = \psi_e(\mathbf{1}), \tag{4.53}$$

and

$$\|\mathbf{F}\| \to \infty \Rightarrow \psi_e(\mathbf{F}) \to 0. \tag{4.54}$$

Here ψ_f and $\psi_e(\mathbf{F})$ designate the constant bulk *failure energy* and the *elastic energy* respectively; $H(\alpha)$ is a unit step function, i.e. $H(z) = 0$ if $z < 0$ and $H(z) = 1$ otherwise.

The switch parameter $\alpha \in (-\infty, 0]$ is defined by the evolution equation

$$\dot{\alpha} = -H\left(\epsilon - \frac{\psi_e}{\psi_f}\right), \quad \alpha(t = 0) = 0, \tag{4.55}$$

where $0 < \epsilon \ll 1$ is a dimensionless precision constant.

The physical interpretation of (4.52) is straightforward: material response is hyperelastic as long as the strain energy is below its limit—ψ_f. When the limit is reached, then the strain energy remains constant for the rest of the deformation process, thereby making material the healing impossible. Parameter α is *not an internal variable*. It functions as a switch: if $\alpha = 0$ then the process is elastic and if $\alpha < 0$ then the material is irreversibly damaged and the strain energy is dissipated.

In order to enforce the energy limiter in the strain energy function, we use the following form of the elastic energy, for example,

$$\psi_e(\mathbf{F}) = \frac{\Phi}{m}\Gamma\left(\frac{1}{m}, \frac{W(\mathbf{F})^m}{\Phi^m}\right), \tag{4.56}$$

where $\Gamma(s, x) = \int_x^\infty t^{s-1}e^{-t}dt$ is the upper incomplete gamma function; $W(\mathbf{F})$ is the strain energy of *intact* (without failure) material; Φ is the energy limiter,[1] which is calibrated in macroscopic experiments; and m is a dimensionless material parameter, which controls the sharpness of the transition to material failure on the stress-strain curve. Increasing or decreasing m it is possible to simulate more or less steep ruptures of the internal bonds accordingly.

---

[1] The reader should note that we use the same character Φ for the energy limiter as for the angle in curvilinear coordinates. The difference between them is always clear from the context.

The failure energy is computed as follows

$$\psi_f = \psi_e(1) = \frac{\Phi}{m} \Gamma\left(\frac{1}{m}, \frac{W(1)^m}{\Phi^m}\right). \tag{4.57}$$

Differentiating (4.52) we obtain the constitutive law in the form

$$\mathbf{P} = -H(\alpha)\frac{\partial \psi_e}{\partial \mathbf{F}} = \exp\left(-\frac{W^m}{\Phi^m}\right) H(\alpha)\frac{\partial W}{\partial \mathbf{F}}. \tag{4.58}$$

For the sake of illustration, the Yeoh model mentioned in the previous section for the intact material behavior of natural rubber

$$W = \sum_{k=1}^{3} c_k (I_1 - 3)^k \tag{4.59}$$

can be enhanced with the failure parameters: $m = 10$, and $\Phi = 82.0\,\text{MPa}$.

We should finally note that the switch parameter α is important in the processes including unloading. Otherwise, the step function $H(\alpha)$ can be dropped from (4.52).

4.7   Uniaxial Tension

In the case of *uniaxial tension* of incompressible material ($J = 1$) we have

$$y_1 = \lambda x_1, \quad y_2 = \lambda^{-1/2} x_2, \quad y_3 = \lambda^{-1/2} x_3, \tag{4.60}$$

where λ is the stretch in the direction of tension.

The deformation gradient and the left Cauchy–Green tensor take the following forms accordingly

$$\begin{aligned}
\mathbf{F} &= \text{Grad}\mathbf{y} = \lambda \mathbf{e}_1 \otimes \mathbf{e}_1 + \lambda^{-1/2}(\mathbf{e}_2 \otimes \mathbf{e}_2 + \mathbf{e}_3 \otimes \mathbf{e}_3), \\
\mathbf{B} &= \mathbf{F}\mathbf{F}^{\text{T}} = \lambda^2 \mathbf{e}_1 \otimes \mathbf{e}_1 + \lambda^{-1}(\mathbf{e}_2 \otimes \mathbf{e}_2 + \mathbf{e}_3 \otimes \mathbf{e}_3).
\end{aligned} \tag{4.61}$$

Substituting the latter equation in the constitutive law for isotropic hyperelastic incompressible material (4.39) we get

$$\boldsymbol{\sigma} = \sigma_{11}\mathbf{e}_1 \otimes \mathbf{e}_1 + \sigma_{22}\mathbf{e}_2 \otimes \mathbf{e}_2 + \sigma_{33}\mathbf{e}_3 \otimes \mathbf{e}_3, \tag{4.62}$$

where

$$\begin{aligned}
\sigma_{11} &= -\Pi + 2(\psi_1 + I_1\psi_2)\lambda^2 - 2\psi_2\lambda^4, \\
\sigma_{22} &= -\Pi + 2(\psi_1 + I_1\psi_2)\lambda^{-1} - 2\psi_2\lambda^{-2} = \sigma_{33}.
\end{aligned} \tag{4.63}$$

Fig. 4.2 Cauchy stress
[MPa] versus stretch in
uniaxial tension of natural
rubber: *dashed line*
designates the intact model;
solid line designates the
model with the energy limiter

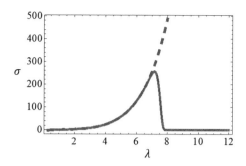

The stresses are homogeneous and the equilibrium equations are satisfied automatically. In uniaxial tension we further assume

$$\sigma_{22} = \sigma_{33} = 0, \tag{4.64}$$

and, consequently,

$$\Pi = 2(\psi_1 + I_1\psi_2)\lambda^{-1} - 2\psi_2\lambda^{-2}. \tag{4.65}$$

Substituting the found Lagrange multiplier Π in the stress tensor we get

$$\sigma = \sigma_{11} = 2(\psi_1 + I_1\psi_2)(\lambda^2 - \lambda^{-1}) + 2\psi_2(\lambda^{-2} - \lambda^4), \tag{4.66}$$

and accounting for

$$I_1 = \lambda^2 + 2\lambda^{-1}, \tag{4.67}$$

we, finally, obtain

$$\sigma = 2(\psi_1 + \psi_2\lambda^{-1})(\lambda^2 - \lambda^{-1}). \tag{4.68}$$

The Cauchy stress–stretch curve for the Yeoh model enhanced with the energy limiter described in the previous section is shown in Fig. 4.2, where also the results are shown for the intact material model. Failure occurs at the limit point at critical stretch $\lambda_{cr} = 7.12$ in accordance with experimental data from Hamdi et al. (2006).

4.8 Biaxial Tension

Biaxial tension tests are often used to calibrate material models. The theoretical background for such tests can be readily developed. Let us consider the homogeneous biaxial deformation of a thin isotropic incompressible sheet of material—Fig. 4.3.

The deformation is controlled by stretches as follows

$$y_1 = \lambda_1 x_1, \quad y_2 = \lambda_2 x_2, \quad y_3 = \lambda_3 x_3. \tag{4.69}$$

By the direct computation we get

Fig. 4.3 Biaxial tension

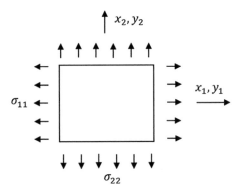

$$\mathbf{F} = \mathrm{Grad}\mathbf{y} = \lambda_1 \mathbf{e}_1 \otimes \mathbf{e}_1 + \lambda_2 \mathbf{e}_2 \otimes \mathbf{e}_2 + \lambda_3 \mathbf{e}_3 \otimes \mathbf{e}_3. \tag{4.70}$$

Thus, the coordinate system coincides with the principal directions of stretches and the constitutive law takes form

$$
\begin{aligned}
\sigma_{11} &= -\Pi + 2(\psi_1 + I_1\psi_2)\lambda_1^2 - 2\psi_2\lambda_1^4, \\
\sigma_{22} &= -\Pi + 2(\psi_1 + I_1\psi_2)\lambda_2^2 - 2\psi_2\lambda_2^4, \\
\sigma_{33} &= -\Pi + 2(\psi_1 + I_1\psi_2)\lambda_3^2 - 2\psi_2\lambda_3^4.
\end{aligned}
\tag{4.71}
$$

The stresses are homogeneous and the equilibrium equations are obeyed automatically.

In biaxial tension we further assume

$$\sigma_{33} = 0, \tag{4.72}$$

and, consequently,

$$\Pi = 2(\psi_1 + I_1\psi_2)\lambda_3^2 - 2\psi_2\lambda_3^4. \tag{4.73}$$

Substituting the Lagrange multiplier Π in the stress tensor we get

$$
\begin{aligned}
\sigma_{11} &= 2(\psi_1 + I_1\psi_2)(\lambda_1^2 - \lambda_3^2) - 2\psi_2(\lambda_1^4 - \lambda_3^4), \\
\sigma_{22} &= 2(\psi_1 + I_1\psi_2)(\lambda_2^2 - \lambda_3^2) - 2\psi_2(\lambda_2^4 - \lambda_3^4).
\end{aligned}
\tag{4.74}
$$

Since

$$I_1 = \mathrm{tr}\mathbf{B} = \lambda_1^2 + \lambda_2^2 + \lambda_3^2, \tag{4.75}$$

we can rewrite stresses in the form

$$
\begin{aligned}
\sigma_{11} &= 2(\lambda_1^2 - \lambda_3^2)(\psi_1 + \psi_2\lambda_2^2), \\
\sigma_{22} &= 2(\lambda_2^2 - \lambda_3^2)(\psi_1 + \psi_2\lambda_1^2),
\end{aligned}
\tag{4.76}
$$

Fig. 4.4 Failure envelope
for biaxial tension of natural
rubber: theory ($\star$) versus
experiment ($\blacktriangle$)

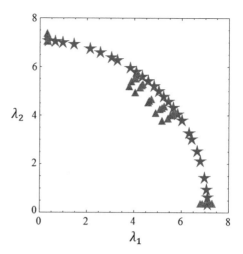

where the incompressibility condition enforces

$$\lambda_3 = \frac{1}{\lambda_1 \lambda_2}. \tag{4.77}$$

Controlling stress and stretch it is possible to fit the strain energy function to the experimental data.

By using the Yeoh model with the energy limiter introduced in Sect. 4.6 it is possible to find the critical failure stretches—the *failure envelope*—by using the critical condition in the form: $(\partial^2 \psi / \partial \lambda_1^2)(\partial^2 \psi / \partial \lambda_2^2) - (\partial^2 \psi / \partial \lambda_1 \partial \lambda_2)^2 = 0$. Comparison of the theory with the experimental data from Hamdi et al. (2006) is shown in Fig. 4.4.

Somewhat lower critical stretches in equal biaxial tension are reasonable in view of the high imperfection sensitivity of the experiments.

4.9   Balloon Inflation

Balloon inflation is another popular deformation and experimental setting used for calibration of soft materials.

Consider the centrally symmetric inflation of a thin spherical membrane—Fig. 4.5.

Its deformation can be described in terms of principal stretches along the directions of the spherical coordinate systems

$$\lambda_\varphi = \lambda_\theta = \frac{2\pi r}{2\pi R} = \frac{r}{R} = \lambda,$$

$$\lambda_r = \frac{h}{H} = \frac{1}{\lambda_\varphi \lambda_\theta} = \lambda^{-2}, \tag{4.78}$$

Fig. 4.5 Inflating balloon

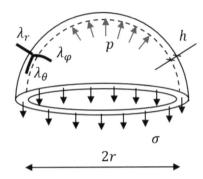

where r, R and h, H are the current and referential radii and thicknesses of the sphere accordingly and incompressibility condition $J = 1$ is taken into account in the second equation.

The deformation gradient and the left Cauchy–Green tensors take the following forms accordingly

$$\mathbf{F} = \lambda^{-2}\mathbf{g}_r \otimes \mathbf{G}_R + \lambda(\mathbf{g}_\theta \otimes \mathbf{G}_\Theta + \mathbf{g}_\varphi \otimes \mathbf{G}_\Phi),$$
$$\mathbf{B} = \mathbf{F}\mathbf{F}^T = \lambda^{-4}\mathbf{g}_r \otimes \mathbf{g}_r + \lambda^2(\mathbf{g}_\theta \otimes \mathbf{g}_\theta + \mathbf{g}_\varphi \otimes \mathbf{g}_\varphi). \tag{4.79}$$

Cauchy stress is

$$\boldsymbol{\sigma} = \sigma_{rr}\mathbf{g}_r \otimes \mathbf{g}_r + \sigma_{\theta\theta}(\mathbf{g}_\theta \otimes \mathbf{g}_\theta + \mathbf{g}_\varphi \otimes \mathbf{g}_\varphi), \tag{4.80}$$

where

$$\sigma_{rr} = -\Pi + 2(\psi_1 + \psi_2 I_1)\lambda^{-4} - 2\psi_2\lambda^{-8},$$
$$\sigma_{\theta\theta} = -\Pi + 2(\psi_1 + \psi_2 I_1)\lambda^2 - 2\psi_2\lambda^4 = \sigma. \tag{4.81}$$

Since the balloon is very thin we assume

$$\sigma_{rr} = 0, \tag{4.82}$$

and, consequently

$$\Pi = 2(\psi_1 + \psi_2 I_1)\lambda^{-4} - 2\psi_2\lambda^{-8}. \tag{4.83}$$

Substituting the found Lagrange multiplier Π back in $\sigma = \sigma_{\theta\theta}$ we get

$$\begin{aligned}
\sigma &= 2(\psi_1 + \psi_2 I_1)\lambda^2(1 - \lambda^{-6}) - 2\psi_2(\lambda^4 - \lambda^{-8}) \\
&= 2\psi_1\lambda^2(1 - \lambda^{-6}) + 2\psi_2[(2\lambda^2 + \lambda^{-4})(\lambda^2 - \lambda^{-4}) - (\lambda^4 - \lambda^{-8})] \\
&= 2\psi_1\lambda^2(1 - \lambda^{-6}) + 2\psi_2\lambda^2(\lambda^2 - \lambda^{-4}) \\
&= 2(\psi_1 + \psi_2\lambda^2)\lambda^2(1 - \lambda^{-6}).
\end{aligned} \tag{4.84}$$

In order to relate stresses to the internal pressure p, we consider equilibrium of a half sphere

$$2\pi r h \sigma = \pi r^2 p,$$ (4.85)

or

$$p = 2\frac{h}{r}\sigma = 2\frac{\lambda^{-2}H}{\lambda R}\sigma = \frac{2H}{\lambda^3 R}\sigma = \frac{4H}{\lambda R}(\psi_1 + \psi_2\lambda^2)(1 - \lambda^{-6}).$$ (4.86)

This analytical formula defines the pressure-stretch curve which is independent of the material choice yet.

In the case of the Yeoh material including failure, which is described in Sects. 4.5 and 4.6, we have

$$\psi = \frac{\Phi}{10}\Gamma\left(\frac{1}{10}, 0\right) - \frac{\Phi}{10}\Gamma\left(\frac{1}{10}, \frac{W^{10}}{\Phi^{10}}\right),$$
$$W = c_1(I_1 - 3) + c_2(I_1 - 3)^2 + c_3(I_1 - 3)^3,$$ (4.87)

where $c_1 = 0.298\,\text{MPa}$, $c_2 = 0.014\,\text{MPa}$, $c_3 = 0.00016\,\text{MPa}$, and $\Phi = 82.0\,\text{MPa}$. Then, we compute the energy derivatives with respect to invariants

$$\psi_1 = \frac{\partial\psi}{\partial I_1} = (c_1 + 2c_2(I_1 - 3) + 3c_3(I_1 - 3)^2)\exp\left(-\frac{W^{10}}{\Phi^{10}}\right), \quad \psi_2 = \frac{\partial\psi}{\partial I_2} = 0.$$ (4.88)

Substitution of the latter equations in (4.86) yields

$$p = \frac{4H}{\lambda R}(1 - \lambda^{-6})(c_1 + 2c_2(I_1 - 3) + 3c_3(I_1 - 3)^2)\exp\left(-\frac{W^{10}}{\Phi^{10}}\right).$$ (4.89)

The pressure-stretch curve defined by this equation is shown in Fig. 4.6.

Qualitatively, the curve exhibits two areas with different slopes. In the beginning the balloon response is stiff and after the critical point of $\lambda \approx 1.3$ it significantly softens and the curve drastically changes its slope. The reader can experience this effect by inflating balloons himself. However, the curve has the limit point at $\lambda \approx 5$ where the material instability and failure occur. The failure description is a result of

Fig. 4.6   Pressure [MPa] versus stretch for inflating balloon made of natural rubber: Yeoh material model with energy limiter

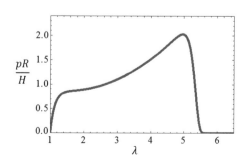

the enforcement of the energy limiter which bounds stresses. The theoretical critical stretch is in a good agreement with the experimental observations presented in Fig. 4.4 for the equibiaxial tension.

4.10   Cavitation

In this section we consider deformation of a very thick sphere under uniform external tension p. When the internal radius of the sphere is very small as compared to the external one it is possible to interpret the problem as the deformation of small *cavity* or void in infinite continuum under *hydrostatic tension*.

Assuming that the deformation is centrally-symmetric and the physical basis vectors in spherical coordinates coincide with the principal directions of stretches, we can write the deformation law as follows

$$r = r(R), \quad \varphi = \Phi, \quad \theta = \Theta. \tag{4.90}$$

Designating the radial direction with index 1 and tangential directions with indices 2 and 3 we can write the principal stretches in the form

$$\lambda_1 = \frac{dr}{dR}, \quad \lambda_2 = \lambda_3 = \frac{r}{R}. \tag{4.91}$$

Since the volume of incompressible material is preserved during deformation we have

$$b^3 - a^3 = B^3 - A^3, \tag{4.92}$$

where A, a and B, b are the internal and external radii of the sphere before and after deformation accordingly. We also notice that any sub-sphere with external radii R, r before and after deformation accordingly also preserves its volume and, consequently, we have

$$r^3 - a^3 = R^3 - A^3. \tag{4.93}$$

The principal components of the Cauchy stress are in the directions of the basis vectors

$$\sigma_1 = \sigma_{rr} = -\Pi + \lambda_1 \frac{\partial \psi}{\partial \lambda_1},$$
$$\sigma_2 = \sigma_{\varphi\varphi} = -\Pi + \lambda_2 \frac{\partial \psi}{\partial \lambda_2}, \tag{4.94}$$
$$\sigma_3 = \sigma_{\theta\theta} = -\Pi + \lambda_3 \frac{\partial \psi}{\partial \lambda_3},$$

where Π is the indefinite Lagrange multiplier.

The stresses should obey the only scalar equilibrium equation

$$\frac{d\sigma_{rr}}{dr} + 2\frac{\sigma_{rr} - \sigma_{\varphi\varphi}}{r} = 0. \tag{4.95}$$

This equation can be integrated as follows

$$\sigma_{rr}(b) - \sigma_{rr}(a) = 2\int_a^b \frac{\sigma_{\varphi\varphi} - \sigma_{rr}}{r} dr, \tag{4.96}$$

or

$$p = 2\int_a^b \left(\lambda_2 \frac{\partial\psi}{\partial\lambda_2} - \lambda_1 \frac{\partial\psi}{\partial\lambda_1}\right) \frac{dr}{r}, \tag{4.97}$$

where boundary conditions are taken into account

$$\sigma_{rr}(a) = 0,$$
$$\sigma_{rr}(b) = p. \tag{4.98}$$

We note that we use the Eulerian description of deformation and

$$R(r, a) = \sqrt[3]{r^3 - a^3 + A^3}. \tag{4.99}$$

To make the formulation dimensionless with respect to length we rewrite (4.97) as follows

$$p = 2\int_{\bar{a}}^{\bar{b}} \left(\lambda_2 \frac{\partial\psi}{\partial\lambda_2} - \lambda_1 \frac{\partial\psi}{\partial\lambda_1}\right) \frac{d\bar{r}}{\bar{r}}, \tag{4.100}$$

where

$$\lambda_1 = \frac{R^2}{r^2} = \frac{\bar{R}^2}{\bar{r}^2}, \quad \lambda_2 = \lambda_3 = \frac{r}{R} = \frac{\bar{r}}{\bar{R}}, \quad \bar{r} = \frac{r}{A}, \quad \bar{R} = \frac{R}{A}, \quad \bar{a} = \frac{a}{A},$$
$$\bar{b} = \frac{b}{A} = \sqrt[3]{\bar{a}^3 + (B/A)^3 - 1}, \quad \bar{R}(\bar{r}, \bar{a}) = \sqrt[3]{\bar{r}^3 - \bar{a}^3 + 1}. \tag{4.101}$$

For $B \gg A$ we have the problem of the expansion of small cavity in the infinite medium under the remote hydrostatic tension. (In computations it is safe to set $B = 10^3 A$.) The graph determined by (4.100) relates the hydrostatic tension with the void hoop stretch, $\bar{a} = a/A$.

It remains to define a material model via the strain energy function. We consider two materials. The first is the Yeoh material enhanced with the energy limiter described in Sects. 4.5 and 4.6. The hydrostatic tension-stretch curve for this material is shown in Fig. 4.7.

The results show that starting from the hydrostatic tension of ~2.3 MPa the cavity expands unstably it yields. It should not be missed that the unstable yield

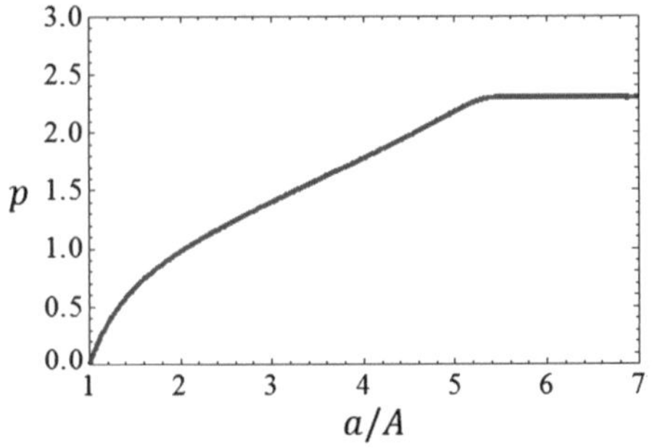

Fig. 4.7 Hydrostatic tension [MPa] versus hoop stretch for cavity growth in natural rubber

Fig. 4.8 Grown visible cavities in the Gent–Lindley poker-chip test

of the cavity—*cavitation*—is a result of the assumption of the centrally-symmetric deformation. This assumption is restrictive, of course, and it can be violated for real inhomogeneous materials. Imperfections can trigger localization of failure in the vicinity of the critical starred point. Nevertheless, the prediction of the critical point of the cavity instability seems to be reasonable even in the presence of imperfections. Moreover, Gent and Lindley (1959) performed "poker-chip" tests on natural rubber specimens and observed the yielding of micron-scale cavities into the visible ones—Fig. 4.8.

Globally, the specimen underwent uniaxial tension while locally the tension was hydrostatic because the specimen was thin and geometrically restrained. The cut, shown in Fig. 4.8, was done at the hydrostatic tension of ∼2.7 MPa. There is an encouraging correlation between the theory and experiment assuming that the Yeoh model is reasonable for the tested sample.

Second material model for cavitation analysis is presented by the strain energy function for the *abdominal aortic aneurysm* (AAA). Aneurysms are abnormal dilatations of vessels in the vascular system. For example, AAA is found in approximately 2 % of men aged 65 years. In many cases AAA gradually expands until rupture causing high mortality rates. The rupture mechanism is not clear. However, the cavitation instability can be a possible trigger of the aneurysm rupture.

We consider the following AAA model (Volokh 2015)

$$\psi = \Phi - \Phi \exp\left(-\frac{c_1(I_1 - 3) + c_2(I_1 - 3)^2}{\Phi}\right), \tag{4.102}$$

where the constants

$$c_1 = 0.52 \text{ MPa}, \quad c_2 = 3.82 \text{ MPa}, \quad \Phi = 0.255 \text{ MPa}$$

were calibrated in uniaxial tension tests shown in Fig. 4.9.

Substituting this strain energy function in (4.100) we can generate the tension-stretch curve shown in Fig. 4.10.

Similarly to the case of cavitation in rubber we have the critical point of the onset of cavitation instability in the AAA material.

At the end of this section we should emphasize that the phenomenon of cavitation instability appears in Figs. 4.7 and 4.10 in the form of yielding because the material models are enhanced with limiters in the strain energy functions. Without the limiters cavitation instabilities would not be observed (Lev and Volokh 2016). Cavitation instability is a material failure phenomenon.

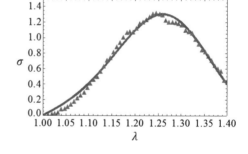

Fig. 4.9 Cauchy stress [MPa] versus stretch for theory and experiment (▲) in uniaxial tension of AAA material

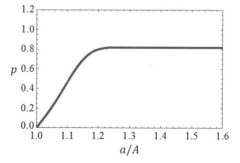

Fig. 4.10 Hydrostatic tension [MPa] versus hoop stretch for cavity growth in AAA material

4.11   On Modeling Crack Propagation

Material failure is a final stage of deformation and all materials fail. In most cases of failure cracks appear and propagate. Continuum mechanics approaches for modeling crack propagation can be provisionally divided in two groups: surface and bulk models.

Surface or *interface failure* models appear by name of *cohesive zone models* in the modern literature. Cohesive zone is a surface in bulk material where displacement discontinuities occur. Thus, continuum is enhanced with discontinuities which require additional traction-separation constitutive equations. The latter equations are constructed qualitatively as follows: traction increases, reaches a maximum, and then tends to zero with the increasing separation. If the location of the cohesive zone is known in advance, as in the case of fracture along weak material interfaces, then the use of the approach is natural. Otherwise, the insertion of the cohesive zone in the bulk, including nucleation, orientation, propagation, branching, and arrest, remains an open problem. Besides, the described approach presumes the use of two different constitutive theories for cohesive zone and bulk for the same real material. Certainly, a correlation between these two constitutive theories is desirable yet not readily available.

Bulk failure models are described by constitutive equations incorporating softening, i.e. descending stress-strain curves. For example, the approach of energy limiters described in Sect. 4.6 is a bulk failure model, which does not introduce *internal variables*. Failure nucleation, propagation, branching, and arrest are a natural outcome of the constitutive formulation including a description of bulk failure. Unfortunately, numerical (finite element) simulations based on the bulk failure models exhibit *pathological mesh-sensitivity*: the finer is the mesh the narrower is the *thickness* of the damage localization zone. Ultimately, the energy dissipated in material failure tends to zero with the decreasing size of the geometric mesh. Such unphysical mesh-sensitivity is a result of the absence of a *characteristic length* in the classical continuum mechanics. To overcome the latter difficulty the *generalized continua* models in the form of *nonlocal* (gradient- or integral- type) continuum failure theories are used, in which a characteristic length that limits the thickness of the failure localization zone is *explicitly* incorporated. The *regularization* strategy based on the introduction of generalized continua is appealing because it is clear and clean mathematically. Unluckily, these generalized continua models are based on the physical assumption of *long-range* (nonlocal) particle interactions while the real particle interactions are *short-range* (on nanometer or, even, angstrom scale). Consequently, the physical basis for the generalized continua models remains debatable. A more physically based approach to handle the pathological mesh-sensitivity of simulations should probably include *multi-physics coupling*. The coupling augments the problem and, by that, naturally regularizes it. Besides, Needleman (1988) argued that the account of rate effects (*viscosity*) might have a regularization effect. It is interesting to note that in both cases of coupled problems and rate effects material length scales are introduced *implicitly*. The latter feature might actually be very attractive because the thickness of the failure localization zone becomes an *outcome* of computations.

In summary, the problem of modeling crack propagation remains largely open.

4.12   Exercises

1. Prove (4.6).
2. Prove (4.10).
3. Prove (4.21).
4. Derive constitutive equations for (4.48).
5. Derive constitutive equations for (4.49).
6. Derive (4.76).

References

Antman SS (1995) Nonlinear problems of elasticity. Springer, Berlin
Barenblatt GI, Joseph DD (eds) (1997) Collected papers of R. S. Rivlin, vol 1–2. Springer, Berlin
Beatty MF (1987) Topics in finite elasticity: hyperelasticity of rubber, elastomers, and biological tissues with examples. Appl Mech Rev 40:1699–1734
Gent AN, Lindley PB (1959) Internal rupture of bonded rubber cylinders in tension. Proc Roy Soc A 2:195–205
Hamdi A, Nait Abdelaziz M, Ait Hocine N, Heuillet P, Benseddiq N (2006) A fracture criterion of rubber-like materials under plane stress conditions. Polym Test 25:994–1005
Holzapfel GA (2000) Nonlinear solid mechanics. Wiley, New York
Lev Y, Volokh KY (2016) On cavitation in rubberlike materials. J Appl Mech 83:044501
Needleman A (1988) Material rate dependence and mesh sensitivity in localization problems. Comp Meth Appl Mech Eng 67:69–85
Needleman A (2014) Some issues in cohesive surface modeling. Procedia IUTAM 10:221–246
Ogden RW (1997) Non-linear elastic deformations. Dover, New York
Sasso M, Palmieri G, Chiappini G, Amodio D (2008) Characterization of hyperelastic rubber-like materials by biaxial and uniaxial stretching tests based on optical methods. Polym Test 27:995–1004
Truesdel C, Noll W (2004) The non-linear field theories of mechanics. Springer, Berlin
Volokh KY (2013) Review of the energy limiters approach to modeling failure of rubber. Rubber Chem Technol 86:470–487
Volokh KY (2014) On irreversibility and dissipation in hyperelasticity with softening. J Appl Mech 81:074501
Volokh KY (2015) Cavitation instability as a trigger of aneurysm rupture. Biomech Model Mechanobiol 14:1071–1079

Chapter 5
Anisotropic Elasticity

Rubber-like materials are usually isotropic. It is possible, of course, to strengthen them by embedding fibers in prescribed directions and creating the *fiber-reinforced composites*. Nature does so with the soft biological tissues which usually consist of an isotropic matrix with the embedded and oriented collagen fibers. The collagen fibers are aligned with the axes of ligaments and tendons forming one characteristic direction or they can form two and more characteristic directions in the case of blood vessels, heart etc.

5.1 On Material Symmetry

In the case of isotropy, material properties are equivalent in all directions. This equivalence can be formalized with the help of the rotated reference configuration. Let point $\mathbf{x}$ be rotated to point $\mathbf{x}'$. Then, the relative deformation gradient is a proper-orthogonal tensor: $\partial \mathbf{x}'/\partial \mathbf{x} = \mathbf{Q}$, where $\mathbf{Q}^{\mathrm{T}} = \mathbf{Q}^{-1}$ and $\det \mathbf{Q} = 1$. The full deformation gradient can be calculated by using the chain rule as follows

$$
\begin{aligned}
\mathbf{F} &= \frac{\partial y_i}{\partial x_j} \mathbf{e}_i \otimes \mathbf{e}_j \\
&= \frac{\partial y_i}{\partial x'_m} \frac{\partial x'_m}{\partial x_j} \mathbf{e}_i \otimes \mathbf{e}_j \\
&= \frac{\partial y_i}{\partial x'_m} \frac{\partial x'_n}{\partial x_j} \delta_{mn} \mathbf{e}_i \otimes \mathbf{e}_j \\
&= \left(\frac{\partial y_i}{\partial x'_m} \mathbf{e}_i \otimes \mathbf{e}_m \right) \left(\frac{\partial x'_n}{\partial x_j} \mathbf{e}_n \otimes \mathbf{e}_j \right)
\end{aligned}
$$

© Springer Science+Business Media Singapore 2016
K. Volokh, *Mechanics of Soft Materials*, DOI 10.1007/978-981-10-1599-1_5

$$= \frac{\partial \mathbf{y}}{\partial \mathbf{x}'} \frac{\partial \mathbf{x}'}{\partial \mathbf{x}},$$
$$= \mathbf{F}'\mathbf{Q}, \tag{5.1}$$

where

$$\mathbf{F}' = \frac{\partial \mathbf{y}}{\partial \mathbf{x}'}, \tag{5.2}$$

and, consequently,

$$\mathbf{F}' = \mathbf{F}\mathbf{Q}^{\mathrm{T}}. \tag{5.3}$$

Now, the symmetry property of the strain energy can be written in the form

$$\psi(\mathbf{F}') = \psi(\mathbf{F}\mathbf{Q}^{\mathrm{T}}) = \psi(\mathbf{F}). \tag{5.4}$$

The set of the proper-orthogonal tensors $\mathbf{Q}$, for which the invariance of the strain energy holds, generates the *symmetry group* of the material with respect to the reference configuration. In the case of isotropy the symmetry group is the full proper orthogonal group, i.e. all rotations.

5.2 Materials with One Characteristic Direction

Materials enjoying one characteristic direction are also called materials with *transverse isotropy*, i.e. isotropy in the planes perpendicular to the preferred direction. Let us designate the preferred direction by unit vector $\mathbf{m}_0$ in the reference configuration. Then, the symmetry group includes rotations obeying condition: $\mathbf{Q}\mathbf{m}_0 = \pm\mathbf{m}_0$; and the strain energy should be invariant in the form

$$\psi(\mathbf{F}\mathbf{Q}^{\mathrm{T}}, \mathbf{Q}\mathbf{m}_0) = \psi(\mathbf{F}, \mathbf{m}_0). \tag{5.5}$$

To meet the latter condition, the strain energy function $\psi(I_1, I_2, I_3, I_4, I_5)$ should depend on two more invariants, as compared to the isotropic case,

$$I_4 = \mathbf{m} \cdot \mathbf{m} = \mathbf{F}\mathbf{m}_0 \cdot \mathbf{F}\mathbf{m}_0 = \mathbf{m}_0 \cdot \mathbf{F}^{\mathrm{T}}\mathbf{F}\mathbf{m}_0 = \mathbf{C} : \mathbf{m}_0 \otimes \mathbf{m}_0,$$
$$I_5 = \mathbf{C}^2 : \mathbf{m}_0 \otimes \mathbf{m}_0, \tag{5.6}$$

where

$$\mathbf{m} = \mathbf{F}\mathbf{m}_0 \tag{5.7}$$

is not a unit vector.

The fourth invariant, I_4, has a clear physical meaning of the squared stretch in the characteristic direction. Tensor product $\mathbf{m}_0 \otimes \mathbf{m}_0$ is often called the *structural* or *structure tensor*, which characterizes the internal design of material.

Differentiating these new invariants with respect to $\mathbf{C}$ we get

$$\frac{\partial I_4}{\partial \mathbf{C}} = \mathbf{m}_0 \otimes \mathbf{m}_0,$$
$$\frac{\partial I_5}{\partial \mathbf{C}} = \mathbf{C}\mathbf{m}_0 \otimes \mathbf{m}_0 + \mathbf{m}_0 \otimes \mathbf{C}\mathbf{m}_0. \tag{5.8}$$

Now, the hyperelastic constitutive law takes form

$$\begin{aligned}
\mathbf{S} &= 2 \sum_{a=1}^{5} \frac{\partial \psi}{\partial I_a} \frac{\partial I_a}{\partial \mathbf{C}} \\
&= 2\{(\psi_1 + I_1 \psi_2)\mathbf{1} - \psi_2 \mathbf{C} + I_3 \psi_3 \mathbf{C}^{-1} \\
&\quad + \psi_4 \mathbf{m}_0 \otimes \mathbf{m}_0 + \psi_5(\mathbf{C}\mathbf{m}_0 \otimes \mathbf{m}_0 + \mathbf{m}_0 \otimes \mathbf{C}\mathbf{m}_0)\},
\end{aligned} \tag{5.9}$$

or, for the Cauchy stress

$$\begin{aligned}
\sigma &= J^{-1}\mathbf{F}\mathbf{S}\mathbf{F}^{\mathrm{T}} \\
&= 2J^{-1}\{(\psi_1 + I_1 \psi_2)\mathbf{B} - \psi_2 \mathbf{B}^2 + I_3 \psi_3 \mathbf{1} \\
&\quad + \psi_4 \mathbf{m} \otimes \mathbf{m} + \psi_5(\mathbf{B}\mathbf{m} \otimes \mathbf{m} + \mathbf{m} \otimes \mathbf{B}\mathbf{m})\},
\end{aligned} \tag{5.10}$$

where $\mathbf{B} = \mathbf{F}\mathbf{F}^{\mathrm{T}}$ is the left Cauchy-Green tensor.

In the case of incompressible material with $I_3 = 1$ we have instead

$$\begin{aligned}
\sigma &= -\Pi\mathbf{1} + 2\{(\psi_1 + I_1 \psi_2)\mathbf{B} - \psi_2 \mathbf{B}^2 \\
&\quad + \psi_4 \mathbf{m} \otimes \mathbf{m} + \psi_5(\mathbf{B}\mathbf{m} \otimes \mathbf{m} + \mathbf{m} \otimes \mathbf{B}\mathbf{m})\},
\end{aligned} \tag{5.11}$$

where Π is the indeterminate Lagrange parameter enforcing incompressibility.

5.3 Materials with Two Characteristic Directions

In the case of two preferred directions we designate the second characteristic unit vector with prime $\mathbf{m}_0'$. The strain energy should obey the invariance condition in the form

$$\psi(\mathbf{F}\mathbf{Q}^{\mathrm{T}}, \mathbf{Q}\mathbf{m}_0, \mathbf{Q}\mathbf{m}_0') = \psi(\mathbf{F}, \mathbf{m}_0, \mathbf{m}_0'). \tag{5.12}$$

To meet the latter condition, the strain energy function $\psi(I_1, I_2, I_3, I_4, I_5, I_6, I_7, I_8)$ should additionally depend on three more invariants

$$I_6 = \mathbf{m}' \cdot \mathbf{m}' = \mathbf{C} : \mathbf{m}'_0 \otimes \mathbf{m}'_0,$$

$$I_7 = \mathbf{C}^2 : \mathbf{m}'_0 \otimes \mathbf{m}'_0, \qquad (5.13)$$

$$I_8 = \mathbf{C} : \mathbf{m}_0 \otimes \mathbf{m}'_0,$$

where

$$\mathbf{m}' = \mathbf{F}\mathbf{m}'_0 \qquad (5.14)$$

is not a unit vector.

Invariants I_6, I_7 are analogous to I_4, I_5 while invariant I_8 is related to both characteristic directions. We note that the latter invariant changes sign when $\mathbf{m}_0$ or $\mathbf{m}'_0$ is reversed. The latter issue should be taken into account in the strain energy expression. It is possible, for example, to use I_8^2 instead of I_8. Some authors suggest using $(\mathbf{m}_0 \cdot \mathbf{m}'_0) I_8$ instead of I_8. The drawback of the latter proposal is that $\mathbf{m}_0 \cdot \mathbf{m}'_0$ vanishes for the orthogonal vectors.

Differentiating the new invariants with respect to $\mathbf{C}$ we get

$$\frac{\partial I_6}{\partial \mathbf{C}} = \mathbf{m}'_0 \otimes \mathbf{m}'_0,$$

$$\frac{\partial I_7}{\partial \mathbf{C}} = \mathbf{C}\mathbf{m}'_0 \otimes \mathbf{m}'_0 + \mathbf{m}'_0 \otimes \mathbf{C}\mathbf{m}'_0, \qquad (5.15)$$

$$\frac{\partial I_8}{\partial \mathbf{C}} = \frac{1}{2}(\mathbf{m}_0 \otimes \mathbf{m}'_0 + \mathbf{m}'_0 \otimes \mathbf{m}_0),$$

where the last derivative preserves symmetry.

Now the constitutive equation for the Cauchy stress takes form

$$\begin{aligned} J\boldsymbol{\sigma} &= 2(\psi_1 + I_1\psi_2)\mathbf{B} - 2\psi_2\mathbf{B}^2 + 2I_3\psi_3\mathbf{1} \\ &+ 2\psi_4\mathbf{m} \otimes \mathbf{m} + 2\psi_5(\mathbf{B}\mathbf{m} \otimes \mathbf{m} + \mathbf{m} \otimes \mathbf{B}\mathbf{m}) \\ &+ 2\psi_6\mathbf{m}' \otimes \mathbf{m}' + 2\psi_7(\mathbf{B}\mathbf{m}' \otimes \mathbf{m}' + \mathbf{m}' \otimes \mathbf{B}\mathbf{m}') \\ &+ \psi_8(\mathbf{m} \otimes \mathbf{m}' + \mathbf{m}' \otimes \mathbf{m}). \end{aligned} \qquad (5.16)$$

In the case of incompressible material we have instead

$$\begin{aligned} \boldsymbol{\sigma} &= -\Pi\mathbf{1} + 2(\psi_1 + I_1\psi_2)\mathbf{B} - 2\psi_2\mathbf{B}^2 \\ &+ 2\psi_4\mathbf{m} \otimes \mathbf{m} + 2\psi_5(\mathbf{B}\mathbf{m} \otimes \mathbf{m} + \mathbf{m} \otimes \mathbf{B}\mathbf{m}) \\ &+ 2\psi_6\mathbf{m}' \otimes \mathbf{m}' + 2\psi_7(\mathbf{B}\mathbf{m}' \otimes \mathbf{m}' + \mathbf{m}' \otimes \mathbf{B}\mathbf{m}') \\ &+ \psi_8(\mathbf{m} \otimes \mathbf{m}' + \mathbf{m}' \otimes \mathbf{m}). \end{aligned} \qquad (5.17)$$

As an example of anisotropic material model with two characteristic directions we consider the *Holzapfel–Gasser–Ogden* (HGO) strain energy, which became very popular for modeling soft biological tissues, especially, arteries

$$\psi = \frac{c_1}{2}(I_1 - 3) + \frac{c_2}{2c_3} \sum_{i=4,6} \{\exp[c_3(I_i - 1)^2] - 1\}, \qquad (5.18)$$

where c_1, c_2, c_3 are material constants and $\mathbf{m}$ and $\mathbf{m}'$ are symmetric with respect to the plane that is perpendicular to the arterial axis.

The first term on the right-hand side of (5.18) describes the neo-Hookean (extra-cellular) matrix and the second term on the right-hand side of (5.18) describes two families of collagen fibers embedded in the matrix. An important feature of this *exponential* model is its ability to mimic two stages of the deformation of soft tissue with collagen fibers. At the first stage the initially wavy fibers straighten and they do not contribute essentially to the matrix response. However, at the second stage the straightened fibers are mobilized and they make the tissue response much stiffer than it was at the first stage.

5.4 Materials with Multiple Characteristic Directions (Fiber Dispersion)

Collagen fibers comprising soft biological tissues are not ideally aligned with some characteristic directions. Their directions are dispersed in space. The idea of modeling fibrous tissues via the strain energy function including the spatial fiber dispersion is due to Lanir (1983). The strain energy function of the dispersed fibers can be written as follows

$$\psi = \int_0^{2\pi} \int_0^{\pi} \rho(\Theta, \Phi) f(\lambda) \sin \Theta \, d\Theta \, d\Phi, \tag{5.19}$$

where the spherical coordinates are used as in Fig. 1.6; $f(\lambda)$ is the strain energy *density* of an individual fiber as a function of its stretch in tension only $\lambda = \sqrt{\mathbf{a} \cdot \mathbf{Ca}} \geq 1$, where the generic fiber direction is

$$\mathbf{a}(\Theta, \Phi) = \cos \Phi \sin \Theta \mathbf{e}_1 + \sin \Phi \sin \Theta \mathbf{e}_2 + \cos \Theta \mathbf{e}_3; \tag{5.20}$$

and $\rho(\Theta, \Phi)$ is the angular density of the fiber distribution normalized as follows

$$\int_0^{2\pi} \int_0^{\pi} \rho(\Theta, \Phi) \sin \Theta \, d\Theta \, d\Phi = 4\pi. \tag{5.21}$$

Differentiating energy function (5.19) with respect to strain we obtain the constitutive law

$$\mathbf{S} = 2\frac{\partial \psi}{\partial \mathbf{C}} = 2 \int_0^{2\pi} \int_0^{\pi} \rho(\Theta, \Phi) \frac{\partial f}{\partial \mathbf{C}} \sin \Theta \, d\Theta \, d\Phi, \tag{5.22}$$

where the derivative of the fiber energy is calculated as follows

$$\frac{\partial f}{\partial \mathbf{C}} = \frac{\partial f}{\partial \lambda} \frac{\partial \lambda}{\partial \mathbf{C}} = \frac{1}{2\lambda} \frac{\partial f}{\partial \lambda} \mathbf{a} \otimes \mathbf{a}. \tag{5.23}$$

With account of the latter equation the constitutive law for the second Piola–Kirchhoff stress takes the final form

$$\mathbf{S} = \int_0^{2\pi} \int_0^{\pi} \rho(\Theta, \Phi) \frac{1}{\lambda} \frac{\partial f}{\partial \lambda} \mathbf{a} \otimes \mathbf{a} \sin \Theta d\Theta d\Phi. \qquad (5.24)$$

While the described approach is physically appealing it has a computational drawback—the necessity to integrate the constitutive law numerically. Such *integration on a unit sphere* is by no means trivial and a number of proposals how to do that exist in the literature. In this regard, the work by Bazant and Oh (1986) should be mentioned in which a few efficient numerical integration schemes are proposed.

In order to circumvent the integration on a unit sphere an alternative approach has been proposed based on the introduction of the *generalized structure tensor* (Freed et al. 2005; Gasser et al. 2006). The idea, in the latter case, is to account for the fiber dispersion in a structure tensor rather than in the strain energy directly. Particularly, the generalized structure tensor is introduced in the form[1]

$$\mathbf{H} = \frac{1}{4\pi} \int_0^{2\pi} \int_0^{\pi} \rho(\Theta, \Phi) \mathbf{a} \otimes \mathbf{a} \sin \Theta d\Theta d\Phi. \qquad (5.25)$$

Note that $\mathrm{tr}\mathbf{H} = 1$ because $|\mathbf{a}| = 1$ and $\rho(\Theta, \Phi)$ is normalized. Moreover, the generalized structure tensor is symmetric $\mathbf{H} = \mathbf{H}^{\mathrm{T}}$ and its Cartesian components can be readily calculated

$$\mathbf{H} = H_{ij} \mathbf{e}_i \otimes \mathbf{e}_j, \qquad (5.26)$$

where

$$\begin{aligned}
H_{11} &= \frac{1}{4\pi} \int_0^{2\pi} \int_0^{\pi} \rho \sin^3 \Theta \cos^2 \Phi d\Theta d\Phi, \\
H_{22} &= \frac{1}{4\pi} \int_0^{2\pi} \int_0^{\pi} \rho \sin^3 \Theta \sin^2 \Phi d\Theta d\Phi, \\
H_{12} &= \frac{1}{4\pi} \int_0^{2\pi} \int_0^{\pi} \rho \sin^3 \Theta \sin \Phi \cos \Phi d\Theta d\Phi, \\
H_{23} &= \frac{1}{4\pi} \int_0^{2\pi} \int_0^{\pi} \rho \sin^2 \Theta \cos \Theta \sin \Phi d\Theta d\Phi, \\
H_{13} &= \frac{1}{4\pi} \int_0^{2\pi} \int_0^{\pi} \rho \sin^2 \Theta \cos \Theta \cos \Phi d\Theta d\Phi, \\
H_{33} &= 1 - H_{11} - H_{22}.
\end{aligned} \qquad (5.27)$$

---

[1] We use the same notation $\mathbf{H}$ for the generalized structure tensor as we used for the displacement gradient in (2.4). The meaning of the notation is usually clear from the context.

The reader should not miss that the numerical integration has to be done only once in order to calculate the components of the generalized structure tensor. Once calculated this tensor does not change during any possible stress analysis for the given material. The latter is in contrast to the energy-based fiber dispersion model (5.19), which must be recomputed constantly during the analyses of deformation processes.

By analogy with Sects. 5.2 and 5.3 we can introduce strain energy $\psi(I_1, I_2, I_3, K_4, K_5)$ as a function of new invariants

$$K_4 = \mathbf{C} : \mathbf{H},$$
$$K_5 = \mathbf{C}^2 : \mathbf{H}. \tag{5.28}$$

The meaning of K_4 is easily clarified by the direct calculation

$$
\begin{aligned}
K_4 &= \frac{1}{4\pi} \int_0^{2\pi} \int_0^{\pi} \rho(\Theta, \Phi) \mathbf{C} : \mathbf{a} \otimes \mathbf{a} \sin \Theta d\Theta d\Phi \\
&= \frac{1}{4\pi} \int_0^{2\pi} \int_0^{\pi} \rho(\Theta, \Phi) \lambda^2 \sin \Theta d\Theta d\Phi,
\end{aligned} \tag{5.29}
$$

which implies the *averaged squared stretch*.

Differentiating the new invariants with respect to $\mathbf{C}$ we get

$$
\begin{aligned}
\frac{\partial K_4}{\partial \mathbf{C}} &= \mathbf{H}, \\
\frac{\partial K_5}{\partial \mathbf{C}} &= \mathbf{HC} + \mathbf{CH}.
\end{aligned} \tag{5.30}
$$

With account of the latter derivatives the hyperelastic constitutive law takes form

$$\mathbf{S} = 2\{(\psi_1 + I_1\psi_2)\mathbf{1} - \psi_2\mathbf{C} + I_3\psi_3\mathbf{C}^{-1} + \frac{\partial \psi}{\partial K_4}\mathbf{H} + \frac{\partial \psi}{\partial K_5}(\mathbf{HC} + \mathbf{CH})\}. \tag{5.31}$$

It should be noted that invariants K_4 and K_5 do not allow to exclude fibers in compression.

Since the generalized structure tensors induce averaged stretches then lower stresses are expected in models based on $\psi(I_1, I_2, I_3, K_4, K_5)$ as compared to models based on (5.19). The difference between both approaches diminishes as long as fibers are stronger aligned with characteristic directions and less dispersed. We also emphasize that the reinforcing fibers are active for $I_4 > 1$ rather than for $K_4 > 1$.

In the important particular case of transverse isotropy, the angular density of the fiber distribution does not depend on angle Φ and it depends on angle Θ only: $\rho(\Theta)$. The fibers are distributed in rotationally symmetric manner about axes x_3 defined by the unit vector $\mathbf{e}_3$. In this case, the normalization condition reduces to

$$\int_0^\pi \rho(\Theta) \sin \Theta d\Theta = 2. \tag{5.32}$$

The reduced generalized structure tensor becomes diagonal

$$\mathbf{H} = \varpi \mathbf{1} + (1 - 3\varpi)\mathbf{e}_3 \otimes \mathbf{e}_3, \tag{5.33}$$

where

$$\varpi = \frac{1}{4} \int_0^\pi \rho(\Theta) \sin^3 \Theta d\Theta \tag{5.34}$$

is a *single* measure of the degree of dispersion.

It is not obligatory, of course, to choose $\mathbf{e}_3$ as the axis of rotational symmetry and any direction $\mathbf{m}_0$ can be chosen. Thus, (5.33) can be rewritten as

$$\mathbf{H} = \varpi \mathbf{1} + (1 - 3\varpi)\mathbf{m}_0 \otimes \mathbf{m}_0. \tag{5.35}$$

Two limit cases of ϖ should be mentioned: $\varpi = 0$ and $\varpi = 1/3$. In the former case we have a perfect fiber alignment without dispersion: $\mathbf{H} = \mathbf{m}_0 \otimes \mathbf{m}_0$. In the latter case all fibers are equally distributed in all directions: $\mathbf{H} = 1/3$.

Calculating the new invariants we obtain for the case of the transverse isotropy

$$\begin{aligned} K_4 &= \varpi I_1 + (1 - 3\varpi)I_4, \\ K_5 &= \varpi(I_1^2 - 2I_2) + (1 - 3\varpi)I_5, \end{aligned} \tag{5.36}$$

where, we remind: $I_1 = \text{tr}\mathbf{C}$; $2I_2 = I_1^2 - \text{tr}\mathbf{C}^2$; $I_4 = \mathbf{C} : \mathbf{m}_0 \otimes \mathbf{m}_0$; and $I_5 = \mathbf{C}^2 : \mathbf{m}_0 \otimes \mathbf{m}_0$.

From the mathematical perspective the use of the generalized structure tensor $\mathbf{H}$ rather than structure tensor $\mathbf{m}_0 \otimes \mathbf{m}_0$ allows increasing the number of material constants by including ϖ into consideration. This adds flexibility to the constitutive law. We note also that there is no need to define the angular density of the fiber distribution in advance in this case because ϖ is fitted in macroscopic experiments.

In spirit of Sect. 5.3 it is possible to introduce the second family of dispersed fibers via characteristic direction $\mathbf{m}_0'$. For example, the popular alternative to HGO strain energy can be produced by replacing I_4 with K_4 and I_6 with K_6 in (5.18) as follows

$$\psi = \frac{c_1}{2}(I_1 - 3) + \frac{c_2}{2c_3} \sum_{i=4,6} \{\exp[c_3(K_i - 1)^2] - 1\}, \tag{5.37}$$

where

$$K_6 = \varpi I_1 + (1 - 3\varpi)I_6. \tag{5.38}$$

This is the *Gasser–Ogden–Holzapfel* (GOH) model.

5.5 Fung's Model of Biological Tissue

The use of *structure tensors* is not the only way to describe anisotropy. Historically first, Yuan-Cheng Fung and his disciples introduced anisotropy by using the Green strain as follows

$$\psi(\mathbf{E}) = \frac{1}{2}\mathbf{E} : \mathbb{Z}_1 : \mathbf{E} + (\mathbf{E} : \mathbb{Z}_2 : \mathbf{E}) \exp(\mathbf{Z} : \mathbf{E} + \mathbf{E} : \mathbb{Z}_3 : \mathbf{E} + \ldots), \qquad (5.39)$$

where $\mathbf{Z}$ and $\mathbb{Z}_a$ are second- and fourth- order tensors of material constants, respectively.

The exponential function allows modeling stiffening typical of soft biological tissues. As an example of the calibrated Fung strain energy we present the constitutive model of a rabbit carotid artery

$$\psi = \frac{c}{2} \left\{ \exp(c_1 E_{RR}^2 + c_2 E_{\Phi\Phi}^2 + c_3 E_{ZZ}^2 + 2c_4 E_{RR} E_{\Phi\Phi} + 2c_5 E_{ZZ} E_{\Phi\Phi} + 2c_6 E_{RR} E_{ZZ}) - 1 \right\}$$

$$(5.40)$$

with dimensional constant $c = 26.95$ KPa and dimensionless constants

$$c_1 = 0.0089, \ c_2 = 0.9925, \ c_3 = 0.4180, \ c_4 = 0.0193, \ c_5 = 0.0749, \ c_6 = 0.0295.$$

5.6 Artery Under Blood Pressure

We consider inflation of an artery under blood pressure. The corresponding boundary value problem (BVP) includes equations of momentum balance (equilibrium) in Ω

$$\operatorname{div}\boldsymbol{\sigma} = \mathbf{0}, \qquad (5.41)$$

constitutive law

$$\boldsymbol{\sigma} = -\Pi \mathbf{1} + \mathbf{F}\frac{\partial\psi}{\partial\mathbf{E}}\mathbf{F}^{\mathrm{T}}, \qquad (5.42)$$

and boundary conditions on placements and tractions on $\partial\Omega$

$$\mathbf{y} = \bar{\mathbf{y}} \quad \text{or} \quad \boldsymbol{\sigma}\mathbf{n} = \bar{\mathbf{t}}, \qquad (5.43)$$

where "div" operator is with respect to the current coordinates $\mathbf{y}$; $\boldsymbol{\sigma}$ is the Cauchy stress tensor; Π is an unknown Lagrange multiplier; $\mathbf{E} = (\mathbf{F}^{\mathrm{T}}\mathbf{F} - \mathbf{1})/2$ is the Green strain tensor; ψ is the strain energy; $\mathbf{t}$ is traction per unit area of the current surface with the unit outward normal $\mathbf{n}$; and the barred quantities are prescribed.

We consider the radial inflation of an artery as a symmetric deformation of an infinite cylinder and assume the deformation law in the form

$$r = \sqrt{\frac{R^2 - A^2}{\gamma s} + a^2}, \quad \varphi = \gamma \Phi, \quad z = sZ, \tag{5.44}$$

where a point occupying position R, Φ, Z in the reference configuration moves to position r, φ, z in the current configuration; s is the axial stretch; $\gamma = 2\pi/(2\pi - \omega)$, where ω is the artery opening angle in the reference configuration; A, B and a, b are the internal and external artery radii before and after deformation accordingly— Fig. 5.1.

The opening central angle, ω, in a stress-free reference configuration is used to represent *residual stresses*, which are one of the most intriguing features of living tissues. The understanding of residual stresses is an open problem of biomechanics, which many researchers relate to the process of *tissue growth*. The best way to quantify residual stresses remains to be settled.

The deformation gradient and the Green strain take the following forms

$$\mathbf{F} = \frac{R}{\gamma sr}\mathbf{g}_r \otimes \mathbf{G}_R + \frac{\gamma r}{R}\mathbf{g}_\varphi \otimes \mathbf{G}_\Phi + s\mathbf{g}_z \otimes \mathbf{G}_Z,$$

$$\mathbf{E} = E_{RR}\mathbf{G}_R \otimes \mathbf{G}_R + E_{\Phi\Phi}\mathbf{G}_\Phi \otimes \mathbf{G}_\Phi + E_{ZZ}\mathbf{G}_Z \otimes \mathbf{G}_Z, \tag{5.45}$$

where

$$E_{RR} = \frac{R^2}{2(\gamma sr)^2} - \frac{1}{2},$$

$$E_{\Phi\Phi} = \frac{(\gamma r)^2}{2R^2} - \frac{1}{2}, \tag{5.46}$$

$$E_{ZZ} = \frac{s^2}{2} - \frac{1}{2}.$$

Here $\mathbf{G}_R$, $\mathbf{G}_\Phi$, $\mathbf{G}_Z$ and $\mathbf{g}_r$, $\mathbf{g}_\varphi$, $\mathbf{g}_z$ are the orthonormal basis vectors in cylindrical coordinates at the reference and current configurations accordingly.

Fig. 5.1 Artery under pressure

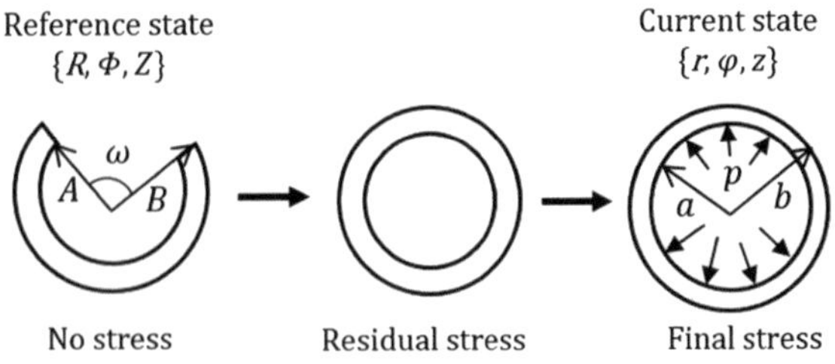

Reference state $\{R, \Phi, Z\}$ Current state $\{r, \varphi, z\}$

No stress   Residual stress   Final stress

With account of the assumed deformation mode we obtain the constitutive law for the following nonzero components of the Cauchy stress

$$\sigma_{rr} = -\Pi + \frac{R^2}{(\gamma sr)^2} \frac{\partial \psi}{\partial E_{RR}},$$

$$\sigma_{\varphi\varphi} = -\Pi + \frac{(\gamma r)^2}{R^2} \frac{\partial \psi}{\partial E_{\Phi\Phi}}, \tag{5.47}$$

$$\sigma_{zz} = -\Pi + s^2 \frac{\partial \psi}{\partial E_{ZZ}}.$$

In the case of axial symmetry for a long tube, there is only one nontrivial scalar equilibrium equation

$$\frac{\partial \sigma_{rr}}{\partial r} + \frac{\sigma_{rr} - \sigma_{\varphi\varphi}}{r} = 0. \tag{5.48}$$

The traction boundary conditions are

$$\sigma_{rr}(a) = -p,$$
$$\sigma_{rr}(b) = 0, \tag{5.49}$$

where p is the internal blood pressure.

We integrate the equilibrium equation over the wall thickness with account of the traction boundary conditions

$$p(a) = -\int_a^{b(a)} (\sigma_{rr} - \sigma_{\varphi\varphi}) \frac{dr}{r} = -\int_a^{b(a)} \left(\frac{R^2}{(\gamma sr)^2} \frac{\partial \psi}{\partial E_{RR}} - \frac{(\gamma r)^2}{R^2} \frac{\partial \psi}{\partial E_{\Phi\Phi}} \right) \frac{dr}{r}, \tag{5.50}$$

where

$$b(a) = \sqrt{a^2 + \frac{B^2 - A^2}{\gamma s}}, \tag{5.51}$$

$$R(r) = \sqrt{\gamma s(r^2 - a^2) + A^2}.$$

Equation (5.50) presents the pressure–radius relationship.

We also introduce dimensionless variables as follows

$$\bar{p} = \frac{p}{c}, \quad \bar{\psi} = \frac{\psi}{c}, \quad \bar{\Pi} = \frac{\Pi}{c}, \quad \bar{r} = \frac{r}{A}, \quad \bar{R} = \frac{R}{A}, \quad \bar{a} = \frac{a}{A}, \quad \bar{b} = \frac{b}{A}, \tag{5.52}$$

where c is the shear modulus.

The dimensionless form of (5.50) is

$$
\bar{p}(\bar{a}) = -\int_{\bar{a}}^{\bar{b}} \left(\frac{\bar{R}^2}{(\gamma s \bar{r})^2} \frac{\partial \bar{\psi}}{\partial E_{RR}} - \frac{(\gamma \bar{r})^2}{\bar{R}^2} \frac{\partial \bar{\psi}}{\partial E_{\Phi\Phi}} \right) \frac{d\bar{r}}{\bar{r}}, \tag{5.53}
$$

where

$$
\bar{b}(\bar{a}) = \sqrt{\bar{a}^2 + \frac{(B/A)^2 - 1}{\gamma s}}, \tag{5.54}
$$

$$
\bar{R}(\bar{r})^2 = \sqrt{\gamma s (\bar{r}^2 - \bar{a}^2) + 1}.
$$

The dimensionless parameter $\bar{\Pi}$ is obtained from the equilibrium equation by integration

$$
\bar{\Pi}(\bar{r}) = \frac{\bar{R}(\bar{r})^2}{(\gamma s \bar{r})^2} \frac{\partial \bar{\psi}}{\partial E_{RR}} + \bar{p}(\bar{a}) + \int_{\bar{a}}^{\bar{r}} \left(\frac{\bar{R}(\zeta)^2}{(\gamma s \zeta)^2} \frac{\partial \bar{\psi}}{\partial E_{RR}}(\zeta) - \frac{(\gamma \zeta)^2}{\bar{R}(\zeta)^2} \frac{\partial \bar{\psi}}{\partial E_{\Phi\Phi}}(\zeta) \right) \frac{d\zeta}{\zeta}, \tag{5.55}
$$

and normalized stresses take form

$$
\bar{\sigma}_{rr} = \frac{\sigma_{rr}}{c} = -\bar{\Pi} + \frac{\bar{R}^2}{(\gamma s \bar{r})^2} \frac{\partial \bar{\psi}}{\partial E_{RR}},
$$

$$
\bar{\sigma}_{\varphi\varphi} = \frac{\sigma_{\varphi\varphi}}{c} = -\bar{\Pi} + \frac{(\gamma \bar{r})^2}{\bar{R}^2} \frac{\partial \bar{\psi}}{\partial E_{\Phi\Phi}}, \tag{5.56}
$$

$$
\bar{\sigma}_{zz} = \frac{\sigma_{zz}}{c} = -\bar{\Pi} + s^2 \frac{\partial \bar{\psi}}{\partial E_{ZZ}}.
$$

Fung's constitutive model (5.40) is used to numerically generate the pressure–radius curves and stresses.

Firstly, we set the referential configuration with $\omega = 0°$ and the internal and external reference radii $A = 0.71$ mm and $B = 1.10$ mm accordingly. The pressure–radius and stress distribution curves are presented in Fig. 5.2.

In Fig. 5.2 stress distribution is shown for normalized pressure $\bar{p} = 0.5$, which corresponds to pressure $p = 13.47$ KPa for the shear modulus $c = 26.95$ KPa.

Secondly, we set a prestressed state with $\omega = 160°$ and the internal and external reference radii $A = 1.43$ mm and $B = 1.82$ mm accordingly. The pressure–radius and stress distribution curves are presented in Fig. 5.3.

In Fig. 5.3 stress distribution is again shown for normalized pressure $\bar{p} = 0.5$, which corresponds to pressure $p = 13.47$ KPa for the shear modulus $c = 26.95$ KPa.

It is remarkable that the distribution of the stress in the tangent direction $\bar{\sigma}_{\varphi\varphi}$ alters significantly under the *residual stress* accounted via the reference opening angle $\omega = 160°$. Indeed, $\bar{\sigma}_{\varphi\varphi}$ varies from 8 to 1 without initial stress (Fig. 5.2) and

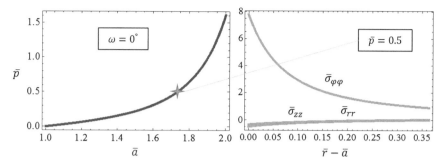

Fig. 5.2 Artery pressure–radius curve without prestress (*left*); and the stress distribution at the normalized pressure equal 0.5 (*right*)

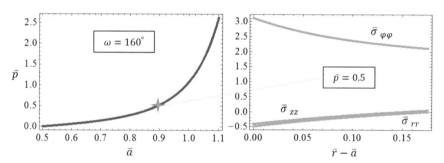

Fig. 5.3 Artery pressure–radius curve with prestress (*left*); and the stress distribution at the normalized pressure equal 0.5 (*right*)

it varies from 3 to 2 in the presence of the initial stress (Fig. 5.3). Nature optimizes the mechanical response of the arterial wall with the help of the residual stresses!

5.7 Exercises

1. Is $C^3 : m_0 \otimes m_0$ independent invariant? Hint: Use the Caley–Hamilton formula (1.45).
2. Prove formulas for the derivatives of invariants $\partial I_4/\partial C$, $\partial I_5/\partial C$, $\partial I_8/\partial C$.
3. Derive constitutive equations for HGO model (5.18).
4. Derive constitutive equations for GOH model (5.37).
5. Derive the arterial deformation law $r = \sqrt{(R^2 - A^2)/(\gamma s) + a^2}$ from the condition of incompressibility.

References

Bazant ZP, Oh BH (1986) Efficient numerical integration on the surface of a sphere. ZAMM 66: 37–49

Cortes DH, Lake SP, Kadlowec JA, Soslowsky LJ, Elliott DM (2010) Characterizing the mechanical contribution of fiber angular distribution in connective tissue: comparison of two modeling approaches. Biomech Model Mechanobiol 9:651–658

Dorfmann L, Ogden RW (eds) (2015) Nonlinear mechanics of soft fibrous materials. Springer, Wien

Freed AD, Einstein DR, Vesely I (2005) Invariant formulation for dispersed transverse isotropy in aortic heart valves. Biomech Model Mechanobiol 4:100–117

Fung YC (1993) Biomechanics: mechanical properties of living tissues. Springer, New York

Gasser TC, Ogden RW, Holzapfel GA (2006) Hyperelastic modeling of arterial layers with distributed collagen fiber orientations. J R Soc Interface 3:15–35

Holzapfel GA, Ogden RW (eds) (2009) Biomechanical modeling at the molecular, cellular and tissue levels. Springer, Wien

Holzapfel GA, Gasser TC, Ogden RW (2000) A new constitutive framework for arterial wall mechanics and a comparative study of material models. J Elast 61:1–48

Lanir Y (1983) Constitutive equations for fibrous connective tissues. J Biomech 16:1–12

Spencer AJM (1984) Continuum theory of the mechanics of fiber-reinforced composites. Springer, Vienna

Volokh KY (2011) Modeling failure of soft anisotropic materials with application to arteries. J Mech Behav Biomed Mater 4:1582–1594

Chapter 6
Incremental Equations

The purpose of the present chapter is to formulate incremental equations of nonlinear elasticity. These equations are useful for analysis of *superimposed small deformations*. Such deformations can tell the story of material stability. A common approach of *linear stability analysis* in continuum mechanics is based on the consideration of growth or decay of superimposed deformations (perturbations).

   We formulate Lagrangean and Eulerian (also called *updated Lagrangean*) initial boundary value problems of nonlinear elasticity with and without the incompressibility constraint. We also consider the propagation of plane waves superimposed on the homogeneous deformations states. An example of the plane wave superimposed on the state of uniaxial tension is elaborated numerically.

6.1 Incremental Equations for Compressible Material

First, we summarize the *initial boundary value problem (IBVP)* of nonlinear elasticity by using the Lagrangean description

$$\rho_0 \frac{\partial^2 \mathbf{y}}{\partial t^2} = \mathrm{Div}\mathbf{P} + \mathbf{b}_0,$$

$$\mathbf{P}\mathbf{F}^{\mathrm{T}} = \mathbf{F}\mathbf{P}^{\mathrm{T}},$$

$$\mathbf{P} = \frac{\partial \psi}{\partial \mathbf{F}}, \tag{6.1}$$

$$\mathbf{P}\mathbf{n}_0 = \bar{\mathbf{t}}_0 \quad \text{or} \quad \mathbf{y} = \bar{\mathbf{y}},$$

$$\mathbf{y}(t = 0) = \mathbf{y}_0,$$

$$\frac{\partial \mathbf{y}}{\partial t}(t = 0) = \mathbf{v}_0,$$

© Springer Science+Business Media Singapore 2016
K. Volokh, *Mechanics of Soft Materials*, DOI 10.1007/978-981-10-1599-1_6

where $\mathbf{y}$ is the placement of a material point; $\mathbf{P}$ is the first Piola-Kirchhoff stress; $\mathbf{b}_0$ is the body force; $\mathbf{F}$ is the deformation gradient; ψ is the strain energy; $\mathbf{n}_0$ is an outward normal to the body surface in the referential configuration; $\bar{\mathbf{t}}_0$ is the prescribed Lagrangean traction on this surface; $\bar{\mathbf{y}}$ is the prescribed surface placement; and $\mathbf{y}_0$ and $\mathbf{v}_0$ are the prescribed initial placement and velocity.

Equations $(6.1)_{1-2}$ are the linear and angular momenta balance in Ω_0; Eq. $(6.1)_3$ is the hyperelastic constitutive law; Eq. $(6.1)_4$ is the natural or essential boundary condition on $\partial \Omega_0$; Eqs. $(6.1)_{5-6}$ are the initial conditions in Ω_0.

Let us *superimpose* small (infinitesimal) increments on placements $\mathbf{y} \rightarrow \mathbf{y} + \tilde{\mathbf{y}}$ where *tilde* designates increment. Then, (6.1) takes form

$$\rho_0 \frac{\partial^2 (\mathbf{y} + \tilde{\mathbf{y}})}{\partial t^2} = \mathrm{Div}(\mathbf{P} + \tilde{\mathbf{P}}) + \mathbf{b}_0,$$

$$(\mathbf{P} + \tilde{\mathbf{P}})(\mathbf{F} + \tilde{\mathbf{F}})^{\mathrm{T}} = (\mathbf{F} + \tilde{\mathbf{F}})(\mathbf{P} + \tilde{\mathbf{P}})^{\mathrm{T}},$$

$$\mathbf{P} + \tilde{\mathbf{P}} = \frac{\partial \psi}{\partial (\mathbf{F} + \tilde{\mathbf{F}})}, \tag{6.2}$$

$$(\mathbf{P} + \tilde{\mathbf{P}})\mathbf{n}_0 = \bar{\mathbf{t}}_0 \quad \text{or} \quad \mathbf{y} + \tilde{\mathbf{y}} = \bar{\mathbf{y}},$$

$$(\mathbf{y} + \tilde{\mathbf{y}})(t = 0) = \mathbf{y}_0,$$

$$\frac{\partial (\mathbf{y} + \tilde{\mathbf{y}})}{\partial t}(t = 0) = \mathbf{v}_0,$$

where

$$\tilde{\mathbf{F}} = \mathrm{Grad}\tilde{\mathbf{y}} = \frac{\partial \tilde{\mathbf{y}}}{\partial \mathbf{x}}. \tag{6.3}$$

We assume here that body force $\mathbf{b}_0$, traction $\bar{\mathbf{t}}_0$, and prescribed placements $\bar{\mathbf{y}}$ are "dead", i.e. they do not depend on deformation.

Subtracting (6.1) from (6.2) and ignoring higher-order incremental terms, we obtain the *incremental IBVP*

$$\rho_0 \frac{\partial^2 \tilde{\mathbf{y}}}{\partial t^2} = \mathrm{Div}\tilde{\mathbf{P}},$$

$$\tilde{\mathbf{P}}\mathbf{F}^{\mathrm{T}} + \mathbf{P}\tilde{\mathbf{F}}^{\mathrm{T}} = (\tilde{\mathbf{P}}\mathbf{F}^{\mathrm{T}} + \mathbf{P}\tilde{\mathbf{F}}^{\mathrm{T}})^{\mathrm{T}},$$

$$\tilde{\mathbf{P}} = \frac{\partial^2 \psi}{\partial \mathbf{F} \partial \mathbf{F}} : \tilde{\mathbf{F}}, \tag{6.4}$$

$$\tilde{\mathbf{P}}\mathbf{n}_0 = \mathbf{0} \quad \text{or} \quad \tilde{\mathbf{y}} = \mathbf{0},$$

$$\tilde{\mathbf{y}}(t = 0) = \mathbf{0},$$

$$\frac{\partial \tilde{\mathbf{y}}}{\partial t}(t = 0) = \mathbf{0},$$

where

$$\frac{\partial^2 \psi}{\partial \mathbf{F} \partial \mathbf{F}} = \frac{\partial^2 \psi}{\partial F_{ij} \partial F_{mn}} \mathbf{e}_i \otimes \mathbf{e}_j \otimes \mathbf{e}_m \otimes \mathbf{e}_n \tag{6.5}$$

is called the fourth-order *elasticity tensor*.

We note that the truncated Taylor series expansion was used for the derivative of the strain energy

$$\frac{\partial \psi}{\partial (\mathbf{F} + \tilde{\mathbf{F}})} = \frac{\partial \psi}{\partial \mathbf{F}} + \frac{\partial^2 \psi}{\partial \mathbf{F} \partial \mathbf{F}} : \tilde{\mathbf{F}}. \tag{6.6}$$

Alternatively, it is possible to reformulate the incremental IBVP in the Eulerian form, in which the current configuration is the referential one. The connection between two descriptions of motion (discussed in Sect. 3.6) is presented by the following formulas

$$\begin{aligned} \rho_0 &= J\rho, \\ \tilde{\mathbf{y}}(\mathbf{x}) &= \tilde{\mathbf{y}}(\mathbf{x}(\mathbf{y})) = \tilde{\mathbf{y}}(\mathbf{y}), \\ \tilde{\mathbf{P}} &= J\tilde{\sigma}\mathbf{F}^{-\mathrm{T}}, \\ \tilde{\mathbf{F}} &= \frac{\partial \tilde{\mathbf{y}}(\mathbf{y})}{\partial \mathbf{y}} \frac{\partial \mathbf{y}}{\partial \mathbf{x}} = \tilde{\mathbf{L}}\mathbf{F}, \\ \mathbf{n}_0 &= \mathbf{F}^{\mathrm{T}}\mathbf{n} \left| \mathbf{F}^{\mathrm{T}}\mathbf{n} \right|^{-1}. \end{aligned} \tag{6.7}$$

Substitution of these formulas in (6.4) yields the incremental IBVP in the Eulerian description[1]

$$\begin{aligned} \rho \frac{\partial^2 \tilde{\mathbf{y}}}{\partial t^2} &= \mathrm{div}\tilde{\sigma}, \\ \tilde{\sigma} + \sigma \tilde{\mathbf{L}}^{\mathrm{T}} &= (\tilde{\sigma} + \sigma \tilde{\mathbf{L}}^{\mathrm{T}})^{\mathrm{T}}, \\ \tilde{\sigma} &= \mathbb{A} : \tilde{\mathbf{L}}, \\ \tilde{\sigma}\mathbf{n} &= \mathbf{0} \quad \text{or} \quad \tilde{\mathbf{y}} = \mathbf{0}, \\ \tilde{\mathbf{y}}(t = 0) &= \mathbf{0}, \\ \frac{\partial \tilde{\mathbf{y}}}{\partial t}(t = 0) &= \mathbf{0}, \end{aligned} \tag{6.8}$$

where $\sigma = J^{-1}\mathbf{P}\mathbf{F}^{\mathrm{T}}$ and the material time derivative is equal to the partial time derivative due to the infinitesimal placement increment.

Equations $(6.8)_{1-2}$ are the linear and angular momenta balance in Ω; Eq. $(6.8)_3$ is the hyperelastic constitutive law; Eq. $(6.8)_4$ is the natural or essential boundary condition on $\partial\Omega$; Eqs. $(6.8)_{5-6}$ are the initial conditions in Ω.

---

[1]We use the blackboard letter $\mathbb{A}$ for the symbolic notation of the fourth-order tensor.

Eulerian elasticity tensor $\mathbb{A}$ is calculated as follows. First, we rewrite $(6.4)_3$ in the form

$$J\tilde{\sigma}\mathbf{F}^{-\mathrm{T}} = \frac{\partial^2 \psi}{\partial \mathbf{F} \partial \mathbf{F}} : (\tilde{\mathbf{L}}\mathbf{F}), \tag{6.9}$$

or

$$\tilde{\sigma} = J^{-1}\{\frac{\partial^2 \psi}{\partial \mathbf{F} \partial \mathbf{F}} : (\tilde{\mathbf{L}}\mathbf{F})\}\mathbf{F}^{\mathrm{T}}. \tag{6.10}$$

Then, by a direct calculation we get

$$\tilde{\sigma}_{ij} = J^{-1}\frac{\partial^2 \psi}{\partial F_{is}\partial F_{km}}\tilde{L}_{kl}F_{lm}F_{js} = A_{ijkl}\tilde{L}_{kl}, \tag{6.11}$$

where

$$A_{ijkl} = J^{-1}F_{js}F_{lm}\frac{\partial^2 \psi}{\partial F_{is}\partial F_{km}} = A_{klij}. \tag{6.12}$$

6.2 Plane Waves in Compressible Material

In this section we analyze the solution of the incremental IBVP in the form of a *plane wave*

$$\tilde{\mathbf{y}}(\mathbf{y}) = \mathbf{r}g(\mathbf{s} \cdot \mathbf{y} - wt), \quad \tilde{y}_i = r_i g(s_k y_k - wt), \tag{6.13}$$

where unit vector $\mathbf{r}$ is called the *wave polarization*; unit vector $\mathbf{s}$ is the direction of the wave propagation; and w is the *wave speed*—see Fig. 6.1.

Using (6.13) and assuming a *homogeneous* state of deformation we calculate

$$\begin{aligned} \frac{\partial^2 \tilde{y}_m}{\partial t^2} &= w^2 r_m g''(s_k y_k - wt), \\ \tilde{L}_{ij} &= \frac{\partial \tilde{y}_i}{\partial y_j} = r_i s_j g'(s_k y_k - wt), \\ \tilde{\sigma}_{mn} &= A_{mnij}r_i s_j g'(s_k y_k - wt), \end{aligned} \tag{6.14}$$

where prime designates differentiation with respect to the argument of function g.

Fig. 6.1 Plane wave

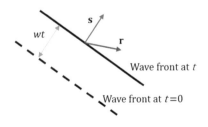

Substituting these formulas in the linear momentum balance $(6.8)_1$ we obtain

$$\rho w^2 r_m = A_{mnij} r_i s_n s_j, \quad \rho w^2 \mathbf{r} = \mathbf{\Lambda}(\mathbf{s})\mathbf{r}, \tag{6.15}$$

where

$$\mathbf{\Lambda}(\mathbf{s}) = \Lambda_{mi} \mathbf{e}_m \otimes \mathbf{e}_i, \quad \Lambda_{mi} = A_{mnij} s_n s_j = \Lambda_{im} \tag{6.16}$$

is called the *acoustic tensor*.

It should not be missed that the entries of the elasticity tensor $\mathbb{A}$ are constants for the assumed homogeneous deformation state. If the state is inhomogeneous then differentiation of $\mathbb{A}$ with respect to spatial coordinates is necessary.

Rewriting the linear momentum balance in the form

$$(\mathbf{\Lambda}(\mathbf{s}) - \rho w^2 \mathbf{1})\mathbf{r} = \mathbf{0}, \tag{6.17}$$

it is easy to recognize the eigenproblem, whose eigenvalues and eigenvectors give the wave speeds and polarization vectors accordingly.

Since the acoustic tensor is symmetric its eigenvalues are real. If all three eigenvalues are positive then material under consideration can propagate three real plane waves.

The scalar product of (6.17) with $\mathbf{r}$ gives

$$\rho w^2 = \mathbf{r} \cdot \mathbf{\Lambda}(\mathbf{s})\mathbf{r} = A_{mnij} s_n s_j r_m r_i, \tag{6.18}$$

and the wave speed is always positive if the acoustic tensor is positive-definite

$$\mathbf{\Lambda}(\mathbf{s}) : \mathbf{r} \otimes \mathbf{r} > 0. \tag{6.19}$$

This is called the *strong ellipticity condition*. If (6.19) holds then all waves are real.

6.3 Incremental Equations for Incompressible Material

Lagrangean incremental IBVP should be modified for incompressible materials

$$\rho_0 \frac{\partial^2 \tilde{\mathbf{y}}}{\partial t^2} = \mathrm{Div}\tilde{\mathbf{P}},$$

$$\tilde{\mathbf{P}}\mathbf{F}^T + \mathbf{P}\tilde{\mathbf{F}}^T = (\tilde{\mathbf{P}}\mathbf{F}^T + \mathbf{P}\tilde{\mathbf{F}}^T)^T,$$

$$\tilde{\mathbf{P}} = \frac{\partial^2 \psi}{\partial \mathbf{F} \partial \mathbf{F}} : \tilde{\mathbf{F}} + \Pi \mathbf{F}^{-T}\tilde{\mathbf{F}}^T\mathbf{F}^{-T} - \tilde{\Pi}\mathbf{F}^{-T},$$

$$\mathrm{tr}(\tilde{\mathbf{F}}\mathbf{F}^{-1}) = 0, \tag{6.20}$$

$$\tilde{\mathbf{P}}\mathbf{n}_0 = \mathbf{0} \quad \text{or} \quad \tilde{\mathbf{y}} = \mathbf{0},$$
$$\tilde{\mathbf{y}}(t = 0) = \mathbf{0},$$
$$\frac{\partial \tilde{\mathbf{y}}}{\partial t}(t = 0) = \mathbf{0},$$

where the incompressibility constraint $\text{tr}(\tilde{\mathbf{F}}\mathbf{F}^{-1}) = 0$ and term $\varPi \mathbf{F}^{-\mathrm{T}}\tilde{\mathbf{F}}^{\mathrm{T}}\mathbf{F}^{-\mathrm{T}} - \tilde{\varPi}\mathbf{F}^{-\mathrm{T}}$ are added to the incremental constitutive law in (6.4).

Eulerian incremental IBVP for incompressible materials takes form

$$\rho \frac{\partial^2 \tilde{\mathbf{y}}}{\partial t^2} = \text{div}\tilde{\boldsymbol{\sigma}},$$
$$\tilde{\boldsymbol{\sigma}} + \boldsymbol{\sigma}\tilde{\mathbf{L}}^{\mathrm{T}} = (\tilde{\boldsymbol{\sigma}} + \boldsymbol{\sigma}\tilde{\mathbf{L}}^{\mathrm{T}})^{\mathrm{T}},$$
$$\tilde{\boldsymbol{\sigma}} = \mathbb{A} : \tilde{\mathbf{L}} + \varPi\tilde{\mathbf{L}}^{\mathrm{T}} - \tilde{\varPi}\mathbf{1}, \tag{6.21}$$
$$\text{tr}\tilde{\mathbf{L}} = 0,$$
$$\tilde{\boldsymbol{\sigma}}\mathbf{n} = \mathbf{0} \quad \text{or} \quad \tilde{\mathbf{y}} = \mathbf{0},$$
$$\tilde{\mathbf{y}}(t = 0) = \mathbf{0},$$
$$\frac{\partial \tilde{\mathbf{y}}}{\partial t}(t = 0) = \mathbf{0},$$

where the incompressibility constraint $\text{tr}\tilde{\mathbf{L}} = 0$ and term $\varPi\tilde{\mathbf{L}}^{\mathrm{T}} - \tilde{\varPi}\mathbf{1}$ are added to the incremental constitutive law in (6.8).

6.4 Plane Waves in Incompressible Material

In the case of the propagation of plane waves in incompressible material we should take the increment of the Lagrange multiplier $\tilde{\varPi}$ into account as follows

$$\tilde{\mathbf{y}}(\mathbf{y}) = \mathbf{r}g(\mathbf{s} \cdot \mathbf{y} - wt),$$
$$\tilde{\varPi} = \Upsilon g'(\mathbf{s} \cdot \mathbf{y} - wt). \tag{6.22}$$

Using (6.22) we calculate

$$\frac{\partial^2 \tilde{y}_m}{\partial t^2} = w^2 r_m g''(s_k y_k - wt),$$
$$\tilde{L}_{ij} = r_i s_j g'(s_k y_k - wt), \tag{6.23}$$
$$\tilde{\sigma}_{mn} = \left(A_{mnij} r_i s_j + \varPi r_n s_m - \Upsilon \delta_{mn}\right) g'(s_k y_k - wt),$$

where prime designates differentiation with respect to the argument of function g.

Substituting these formulas in the linear momentum balance $(6.21)_1$ and incompressibility condition $(6.21)_4$ we obtain respectively

$$\rho w^2 r_m = A_{mnij} r_i s_j s_n + \Pi r_n s_m s_n - \Upsilon \delta_{mn} s_n, \tag{6.24}$$

and

$$r_n s_n = 0. \tag{6.25}$$

These equations can be written in a more compact way

$$\rho w^2 \mathbf{r} = \mathbf{\Lambda}(\mathbf{s})\mathbf{r} - \Upsilon \mathbf{s}, \\ \mathbf{r} \cdot \mathbf{s} = 0. \tag{6.26}$$

Moreover, we can calculate Υ by taking the scalar product of $(6.26)_1$ with $\mathbf{s}$

$$\Upsilon = \mathbf{s} \cdot \mathbf{\Lambda}(\mathbf{s})\mathbf{r}. \tag{6.27}$$

Now, we can rewrite (6.26) in the form

$$\rho w^2 \mathbf{r} = \mathbf{\Lambda}^*(\mathbf{s})\mathbf{r}, \\ \mathbf{r} \cdot \mathbf{s} = 0, \tag{6.28}$$

where

$$\mathbf{\Lambda}^*(\mathbf{s}) = \mathbf{\Lambda}(\mathbf{s}) - \mathbf{s} \otimes \mathbf{\Lambda}(\mathbf{s})\mathbf{s} \tag{6.29}$$

is the modified acoustic tensor for incompressible material.

We note that $\mathbf{\Lambda}^*(\mathbf{s})$ is not symmetric. Moreover, it is singular because zero eigenvalue corresponds to the left eigenvector $\mathbf{s}$

$$\mathbf{\Lambda}^{*T}\mathbf{s} = (\mathbf{\Lambda} - \mathbf{s} \otimes \mathbf{\Lambda}\mathbf{s})^T \mathbf{s} = \mathbf{\Lambda}^T\mathbf{s} - \mathbf{\Lambda}\mathbf{s} = (\mathbf{\Lambda}^T - \mathbf{\Lambda})\mathbf{s} = \mathbf{0}. \tag{6.30}$$

Thus, at most, there are two real plane waves and both them are *transverse* because of the incompressibility condition $(6.28)_2$.

The scalar product of $(6.26)_1$ with $\mathbf{r}$ again provides the real wave speeds for the strong ellipticity condition

$$\rho w^2 = \mathbf{r} \cdot \mathbf{\Lambda}(\mathbf{s})\mathbf{r} - \Upsilon \mathbf{r} \cdot \mathbf{s} = \mathbf{\Lambda}(\mathbf{s}) : \mathbf{r} \otimes \mathbf{r} > 0. \tag{6.31}$$

6.5 Plane Waves Superimposed on Uniaxial Tension

In this section, we consider propagation of plane waves superimposed on the uniaxial tension in incompressible materials (Sect. 4.7). We focus on the wave propagating in the direction of tension

$$\mathbf{s} = \mathbf{e}_1. \tag{6.32}$$

In view of the incompressibility condition $(6.28)_2$ the possible transverse waves are polarized in direction $\mathbf{e}_2$ or $\mathbf{e}_3$. Let us assume

$$\mathbf{r} = \mathbf{e}_2. \tag{6.33}$$

Then, we have

$$\rho w^2 = \mathbf{\Lambda}(\mathbf{e}_1) : \mathbf{e}_2 \otimes \mathbf{e}_2 = \Lambda_{22}(\mathbf{e}_1) = A_{2121} = F_{1s} F_{1m} \frac{\partial^2 \psi}{\partial F_{2s} \partial F_{2m}}. \tag{6.34}$$

In the case of uniaxial tension we have only diagonal nonzero components of the deformation gradient

$$F_{11} = \lambda, \quad F_{22} = \lambda^{-1/2}, \quad F_{33} = \lambda^{-1/2}, \tag{6.35}$$

where λ is the stretch in the direction of tension.

Thus, the formula for the wave speed can be further reduced to

$$\rho w^2 = F_{11} F_{11} \frac{\partial^2 \psi}{\partial F_{21} \partial F_{21}} = \lambda^2 \frac{\partial^2 \psi}{\partial F_{21} \partial F_{21}}. \tag{6.36}$$

Let us restrict the strain energy to the function $\psi(I_1)$ depending on the first invariant $I_1 = \mathbf{F} : \mathbf{F}$ only. Then, we have

$$\frac{\partial \psi}{\partial F_{21}} = \frac{\partial \psi}{\partial I_1} \frac{\partial I_1}{\partial F_{21}} = 2 \frac{\partial \psi}{\partial I_1} F_{21}, \tag{6.37}$$

Fig. 6.2 Speed of transverse superimposed wave [m/s] for uniaxial tension of intact Yeoh material (*dashed line*) and Yeoh material enhanced with the energy limiter (*solid line*)

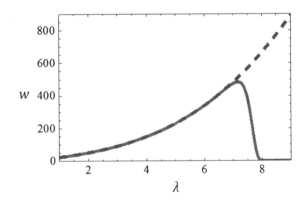

and

$$\frac{\partial^2 \psi}{\partial F_{21} \partial F_{21}} = 4\frac{\partial^2 \psi}{\partial I_1^2}F_{21}^2 + 2\frac{\partial \psi}{\partial I_1} = 2\frac{\partial \psi}{\partial I_1}. \tag{6.38}$$

Thus, the wave speed is

$$\rho w^2 = 2\lambda^2 \psi_1. \tag{6.39}$$

The graph of $w = \sqrt{2\lambda^2 \psi_1/\rho}$, where $\rho = 1100 \text{ kg/m}^3$, is shown in Fig. 6.2 for the case of Yeoh material (4.49).

We present results for the intact Yeoh material and for the Yeoh material enhanced with the energy limiter (Sect. 4.6). It is remarkable that the speed of the superimposed wave increases soundly with stretching. The limit point on the graph corresponds to the limit point in Fig. 4.2 where material failure starts. The onset of material failure makes the wave propagation impossible. The latter conclusion is physically meaningful and appealing.

6.6   Exercises

1. Derive (6.21) from (6.20).

References

Barenblatt GI, Joseph DD (eds) (1997) Collected papers of R. S. Rivlin, vol 1–2. Springer, Berlin
Bigoni D (2012) Nonlinear solid mechanics: bifurcation theory and material instability. Cambridge University Press, Cambridge
Destrade M, Saccomandi G (eds) (2007) Waves in nonlinear pre-stressed materials. Springer, Berlin
Ogden RW (1997) Non-linear elastic deformations. Dover, New York

Chapter 7
Thermoelasticity

The process of mechanical deformation is accompanied or triggered by *heating* or *cooling*. The latter phenomenon is associated with the *thermal* energy in contrast to the purely *mechanical* energy stored in the deformed material that was considered in the previous chapters. The coupled theory called *continuum thermomechanics* or *continuum thermodynamics* has a long history yet its foundations are still open to debate. We do not participate in the debate, however.[1] Instead, we consider the simplest formulation that couples thermal and mechanical processes in the course of deformation of soft materials. Besides, we should note that principles of thermodynamics, especially the *entropy* restrictions, might be important for the constitutive modeling of materials even in the case where the thermal effects are negligible.

7.1 Energy Balance

The *energy balance* or the *first law of thermodynamics* can be written for the whole body or any part of it in the following form

$$\frac{d}{dt} \int \left(\frac{\rho}{2} \mathbf{v} \cdot \mathbf{v} + e \right) dV = \int (\mathbf{v} \cdot \mathbf{b} + r) dV + \oint (\mathbf{v} \cdot \mathbf{t} - \mathbf{q} \cdot \mathbf{n}) dA, \qquad (7.1)$$

where $\mathbf{v}, \rho, \mathbf{b}, \mathbf{t}$ are the velocity, mass density, body force, and traction respectively; e is the *internal energy* per unit current volume (its existence is postulated in this law); $\rho \mathbf{v} \cdot \mathbf{v}/2$ is the *kinetic energy*; r is the rate of the *heat supply* per unit current volume; $-\mathbf{q}$ is a vector of the *heat flux* per unit current area (it is negative because the incoming heat is positive).

---

[1] We regard *energy, entropy, heat flux, temperature* as *primitive* (non-reducible) objects.

© Springer Science+Business Media Singapore 2016
K. Volokh, *Mechanics of Soft Materials*, DOI 10.1007/978-981-10-1599-1_7

The heat is related microscopically to the molecular and atomic fluctuations. Differentiating the integral and using the stress tensor $\mathbf{t} = \boldsymbol{\sigma}\mathbf{n}$, we get

$$\int \left\{ \frac{d}{dt} \left(\frac{\rho}{2}\mathbf{v} \cdot \mathbf{v} + e \right) + \left(\frac{\rho}{2}\mathbf{v} \cdot \mathbf{v} + e \right) \mathrm{div}\mathbf{v} \right\} dV$$
$$= \int (\mathbf{v} \cdot \mathbf{b} + r)dV + \oint (\mathbf{v} \cdot \boldsymbol{\sigma}\mathbf{n} - \mathbf{q} \cdot \mathbf{n})dA. \qquad (7.2)$$

By using the divergence theorem we localize

$$\frac{d}{dt} \left(\frac{\rho}{2}\mathbf{v} \cdot \mathbf{v} + e \right) + \left(\frac{\rho}{2}\mathbf{v} \cdot \mathbf{v} + e \right) \mathrm{div}\mathbf{v} - (\mathbf{v} \cdot \mathbf{b} + r) - \mathrm{div}(\boldsymbol{\sigma}^{\mathrm{T}}\mathbf{v} - \mathbf{q}) = 0, \quad (7.3)$$

where

$$\frac{d}{dt} \left(\frac{\rho}{2}\mathbf{v} \cdot \mathbf{v} \right) + \frac{\rho}{2}\mathbf{v} \cdot \mathbf{v}\,\mathrm{div}\mathbf{v} = \mathbf{v} \cdot \left(\frac{1}{2}\frac{d(\rho\mathbf{v})}{dt} + \frac{\rho}{2}\frac{d\mathbf{v}}{dt} + \frac{\rho}{2}\mathbf{v}\,\mathrm{div}\mathbf{v} \right)$$
$$= \mathbf{v} \cdot \left(\frac{d(\rho\mathbf{v})}{dt} + \rho\mathbf{v}\,\mathrm{div}\mathbf{v} + \frac{\rho}{2}\frac{d\mathbf{v}}{dt} - \frac{1}{2}\frac{d(\rho\mathbf{v})}{dt} - \frac{\rho}{2}\mathbf{v}\,\mathrm{div}\mathbf{v} \right)$$
$$= \mathbf{v} \cdot \left(\frac{d(\rho\mathbf{v})}{dt} + \rho\mathbf{v}\,\mathrm{div}\mathbf{v} \right) - \frac{1}{2}\mathbf{v} \cdot \mathbf{v} \left(\frac{d\rho}{dt} + \rho\,\mathrm{div}\mathbf{v} \right), \qquad (7.4)$$

and

$$\mathrm{div}(\boldsymbol{\sigma}^{\mathrm{T}}\mathbf{v}) = \frac{\partial(\sigma_{mi}v_m)}{\partial y_i} = v_m\frac{\partial\sigma_{mi}}{\partial y_i} + \sigma_{mi}\frac{\partial v_m}{\partial y_i} = \mathbf{v} \cdot \mathrm{div}\boldsymbol{\sigma} + \boldsymbol{\sigma} : \mathbf{L}. \qquad (7.5)$$

By virtue of the linear momentum balance (3.26) and mass conservation (3.5) we obtain finally

$$\dot{e} + e\,\mathrm{div}\mathbf{v} = \boldsymbol{\sigma} : \mathbf{D} + r - \mathrm{div}\mathbf{q}. \qquad (7.6)$$

Lagrangean description of the energy balance takes the following integral form

$$\frac{d}{dt} \int \left(\frac{\rho_0}{2}\mathbf{v} \cdot \mathbf{v} + e_0 \right) dV_0 = \int (\mathbf{v} \cdot \mathbf{b}_0 + r_0)dV_0 + \oint (\mathbf{v} \cdot \mathbf{t}_0 - \mathbf{q}_0 \cdot \mathbf{n}_0)dA_0. \quad (7.7)$$

After localization, it reads

$$\dot{e}_0 = \mathbf{P} : \dot{\mathbf{F}} + r_0 - \mathrm{Div}\mathbf{q}_0, \qquad (7.8)$$

where (see Sect. 3.6)

$$\rho_0 = J\rho, \quad e_0 = Je, \quad r_0 = Jr, \quad \mathbf{t}_0 dA_0 = \mathbf{t}dA,$$
$$\mathbf{b}_0 = J\mathbf{b}, \quad \mathbf{q}_0 = J\mathbf{F}^{-1}\mathbf{q}, \quad \mathbf{n}_0 dA_0 = J^{-1}\mathbf{F}^{\mathrm{T}}\mathbf{n}dA. \qquad (7.9)$$

Thus, the internal energy changes by virtue of (a) stress working on strains (stress power) and (b) volumetric and surface heat supply (thermal power).

We note, finally, that the energy balance on the body surface $\partial\Omega$ reads

$$-\mathbf{q}\cdot\mathbf{n}=\bar{q}, \qquad (7.10)$$

for the Eulerian description or on the body surface $\partial\Omega_0$

$$-\mathbf{q}_0\cdot\mathbf{n}_0=\bar{q}_0, \qquad (7.11)$$

for the Lagrangean description, where $\bar{q}$ or $\bar{q}_0$ is a prescribed heat flux through the unit area of the current or reference surface respectively. The energy balance on the surface presents the *natural* boundary condition.

7.2 Entropy Inequality

According to the first law of thermodynamics the energy does not appear or disappear but rather transforms from one form to another. This law, however, does not account for the *irreversibility* or *direction-dependence* of some physical processes: the heat always flows from the hot body to the cold one and never in the reverse order; friction transforms into heat and never in the reverse order etc.

Entropy is introduced in order to account for the irreversibility of the process. It may be called the degree of *disorder*. It never decreases. It is constant or it increases and *chaos* increases.

Some intuition concerning the concept of entropy can be developed in the case of gas. Figure 7.1 shows a box divided by a wall and filled with gas in one part. When the wall is elevated then the gas tends to spread all over the box - it tends to the state of the maximum disorder. A more ordered state can be imagined in which the gas still fills only one part of the box. The latter would be amazing yet impossible to observe.

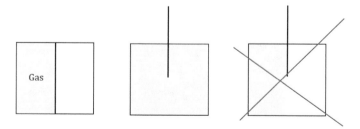

Fig. 7.1 Entropy in gases

The *second law of thermodynamics* is the law of the *entropy production* and it can be written in the following integral form for the whole body or any part of it

$$\frac{d}{dt}\int \eta dV \geq \int \frac{r}{T} dV - \oint \frac{1}{T}\mathbf{q}\cdot\mathbf{n}dA, \qquad (7.12)$$

where η is the *entropy* per unit current volume; r/T is the rate of *entropy supply* per unit current volume and $T \geq 0$ is the *absolute temperature* measured in degrees Kelvin; $\mathbf{q}/T$ is the *entropy flux* per unit current area.

The reader can consider the expressions for the volumetric (r/T) and surface $(\mathbf{q}/T)$ entropy supplies as constitutive laws. Thus, the second law of thermodynamics essentially states that the rate of the entropy alteration is always greater than the entropy supply via heat for irreversible processes. In the case of the reversible process the entropy alteration is equal to the entropy supply.

Localizing the second law of thermodynamics (7.12) we get

$$\dot{\eta} + \eta \mathrm{div}\mathbf{v} \geq \frac{r}{T} - \mathrm{div}\frac{\mathbf{q}}{T}. \qquad (7.13)$$

Substituting for $(r - \mathrm{div}\mathbf{q})$ from the energy balance law (7.6) we obtain the *Clausius–Duhem inequality*

$$\dot{\eta} + \eta \mathrm{div}\mathbf{v} \geq \frac{1}{T}(\dot{e} + e\mathrm{div}\mathbf{v} - \boldsymbol{\sigma} : \mathbf{D}) + \frac{1}{T^2}\mathbf{q}\cdot\mathrm{grad}T. \qquad (7.14)$$

Since the heat flows in the direction of the lower temperature, we have

$$\mathbf{q}\cdot\mathrm{grad}T \leq 0. \qquad (7.15)$$

Thus, we can obey the Clausius–Duhem inequality by imposing a more stringent *Clausius–Planck inequality*

$$\dot{\eta} + \eta \mathrm{div}\mathbf{v} \geq \frac{1}{T}(\dot{e} + e\mathrm{div}\mathbf{v} - \boldsymbol{\sigma} : \mathbf{D}). \qquad (7.16)$$

Lagrangean formulation of the second law of thermodynamics in the integral form is

$$\frac{d}{dt}\int \eta_0 dV_0 \geq \int \frac{r_0}{T} dV_0 - \oint \frac{1}{T}\mathbf{q}_0\cdot\mathbf{n}_0 dA_0,$$

where

$$\eta_0 = J\eta. \qquad (7.17)$$

Its localized Clausius–Duhem form is

$$\dot{\eta}_0 \geq \frac{1}{T}(\dot{e}_0 - \mathbf{P} : \dot{\mathbf{F}}) + \frac{1}{T^2}\mathbf{q}_0\cdot\mathrm{Grad}T. \qquad (7.18)$$

The reduced Clausius–Planck inequality reads

$$\dot{\eta}_0 \geq \frac{1}{T}(\dot{e}_0 - \mathbf{P} : \dot{\mathbf{F}}). \tag{7.19}$$

Rearranging terms in the last inequality we can write down the second law of thermodynamics in terms of the nonnegative *internal dissipation*

$$D_{\text{int}} = \mathbf{P} : \dot{\mathbf{F}} - \dot{e}_0 + T\dot{\eta}_0 \geq 0. \tag{7.20}$$

7.3 Helmholtz's Free Energy

In order to generalize the strain energy function introduced in the previous chapters and enforce the thermo-mechanical coupling it is convenient to define the *Helmholtz free energy* function in the form

$$\psi(\mathbf{F}, T) = e_0 - T\eta_0. \tag{7.21}$$

Differentiating this equation with respect to time we obtain

$$\frac{\partial \psi}{\partial \mathbf{F}} : \dot{\mathbf{F}} + \frac{\partial \psi}{\partial T}\dot{T} = \dot{e}_0 - \dot{T}\eta_0 - T\dot{\eta}_0. \tag{7.22}$$

First, we observe that the substitution of the rate of internal energy from (7.22) into (7.20) yields the second law of thermodynamics in the form

$$D_{\text{int}} = \left(\mathbf{P} - \frac{\partial \psi}{\partial \mathbf{F}}\right) : \dot{\mathbf{F}} - \left(\eta_0 + \frac{\partial \psi}{\partial T}\right)\dot{T} \geq 0. \tag{7.23}$$

The latter law can be obeyed by the following specific choice of the constitutive laws

$$\mathbf{P} = \frac{\partial \psi}{\partial \mathbf{F}},$$
$$\eta_0 = -\frac{\partial \psi}{\partial T}. \tag{7.24}$$

Such choice is usually called the *Coleman–Noll procedure*.

Second, we observe that the substitution of the rate of the internal energy from (7.22) into (7.8) yields the first law of thermodynamics in the form

$$T\dot{\eta}_0 = \left(\mathbf{P} - \frac{\partial \psi}{\partial \mathbf{F}}\right) : \dot{\mathbf{F}} - \left(\eta_0 + \frac{\partial \psi}{\partial T}\right)\dot{T} + r_0 - \text{Div}\mathbf{q}_0, \tag{7.25}$$

which, by virtue of the specified constitutive laws, reduces to

$$T \dot{\eta}_0 = r_0 - \text{Div} \mathbf{q}_0, \tag{7.26}$$

or

$$T \left(-\frac{\partial^2 \psi}{\partial T \partial \mathbf{F}} : \dot{\mathbf{F}} - \frac{\partial^2 \psi}{\partial T^2} \dot{T} \right) = r_0 - \text{Div} \mathbf{q}_0. \tag{7.27}$$

The latter equation can be finally rewritten in the canonical form

$$\zeta \dot{T} = -\text{Div} \mathbf{q}_0 + T \frac{\partial^2 \psi}{\partial T \partial \mathbf{F}} : \dot{\mathbf{F}} + r_0, \tag{7.28}$$

where

$$\zeta = -T \frac{\partial^2 \psi}{\partial T^2} > 0 \tag{7.29}$$

is the *heat capacity*.

We note that the second term on the right hand side of (7.28) provides coupling between mechanical and thermal processes.

The energy balance law (7.28) is completed with the *natural* boundary condition on the body surface $\partial \Omega_0$

$$- \mathbf{q}_0 \cdot \mathbf{n}_0 = \bar{q}_0, \tag{7.30}$$

or the *essential* boundary condition imposed on the temperature

$$T = \bar{T}, \tag{7.31}$$

where $\bar{T}$ is prescribed on $\partial \Omega_0$.

Initial condition on the temperature in Ω_0 is necessary for the time-dependent processes

$$T(t = 0) = T_0, \tag{7.32}$$

where T_0 is prescribed in Ω_0.

The described initial-boundary-value problem is not completely defined yet and it requires a constitutive law for the heat flux. We discuss it in the next section.

7.4 Fourier's Law of Heat Conduction

The first theory of heat conduction, which is essentially a constitutive law, is attributed to Fourier and it can be written in a slightly generalized form as follows

$$\mathbf{q} = -\kappa \text{grad} T, \tag{7.33}$$

where κ is a symmetric *spatial thermal conductivity tensor*.

Substitution of κ in the heat flow restriction (7.15) yields

$$\mathrm{grad}\,T \cdot \kappa\,\mathrm{grad}\,T \geq 0, \tag{7.34}$$

which means that κ must be positive semi-definite.

Original Fourier proposal was

$$\kappa = k\mathbf{1}, \tag{7.35}$$

where $k \geq 0$ is the *coefficient of thermal conductivity*.

Substituting Lagrangean quantities in the heat conduction equation we obtain

$$J^{-1}\mathbf{F}\mathbf{q}_0 = -\kappa\mathbf{F}^{-\mathrm{T}}\mathrm{Grad}\,T, \tag{7.36}$$

or

$$\mathbf{q}_0 = -J\mathbf{F}^{-1}\kappa\mathbf{F}^{-\mathrm{T}}\mathrm{Grad}\,T. \tag{7.37}$$

With account of the heat conduction equation the initial-boundary-value problem of thermo-elasticity is completed.

7.5 Thermoelastic Incompressibility

Many soft materials are nearly incompressible and the condition of incompressibility $J = \det \mathbf{F} = 1$ is often successfully used in analytical solutions of particular problems. In the case of thermoelastic coupling the condition of incompressibility should be modified to account for possible volume changes caused by thermal processes. The assumption of *thermoelastic incompressibility* can be written in the following form

$$\begin{aligned} J &= f(T), \\ f(T_0) &= 1, \end{aligned} \tag{7.38}$$

which means that material is incompressible in the reference state and all volumetric changes are produced by the temperature alterations.

The time increment of this constraint is

$$\frac{\partial f}{\partial T}\dot{T} - J\mathbf{F}^{-\mathrm{T}} : \dot{\mathbf{F}} = 0. \tag{7.39}$$

Here multipliers $\partial f/\partial T$ and $-J\mathbf{F}^{-\mathrm{T}}$ represent workless entropy and stress accordingly, which can be scaled by arbitrary factor Π. Thus, we modify constitutive equations as follows

$$\mathbf{P} = \frac{\partial \psi}{\partial \mathbf{F}} - J\mathbf{F}^{-T}\Pi,$$

$$\eta_0 = -\frac{\partial \psi}{\partial T} + \frac{\partial f}{\partial T}\Pi. \tag{7.40}$$

7.6   Uniaxial Tension

In this section we consider an example of thermoelastic uniaxial tension. We start by specializing the Helmholtz free energy for a rubberlike material in the form

$$\psi = \frac{T}{T_0}\sum_{k=1}^{3}c_k(I_1 - 3)^k + c_0\left(T - T_0 - T\ln\frac{T}{T_0}\right), \tag{7.41}$$

where $c_1 = 0.298\,\text{MPa}$, $c_2 = 0.014\,\text{MPa}$, $c_3 = 0.00016\,\text{MPa}$ and c_0 is a positive constant.

Thus, the constitutive equations for thermoelastically incompressible material take forms

$$\mathbf{P} = 2\frac{T}{T_0}(c_1 + 2c_2(I_1 - 3) + 3c_3(I_1 - 3)^2)\mathbf{F} - J\mathbf{F}^{-T}\Pi,$$

$$\eta_0 = \frac{1}{T_0}\sum_{k=1}^{3}c_k(I_1 - 3)^k - c_0\ln\frac{T}{T_0} + \frac{\partial g}{\partial T}\Pi. \tag{7.42}$$

We further assume uniaxial tension without heat sources ($r_0 = 0$) and with spatially homogeneous temperature and deformations fields. The latter notion has strong implications. Indeed, in the absence of the spatial gradient of the temperature field there is no heat flux

$$\mathbf{q}_0 = \mathbf{0}. \tag{7.43}$$

In this case, the energy balance equation (7.26) reduces to

$$\dot{\eta} = 0 \Rightarrow \eta = \text{constant}. \tag{7.44}$$

Assuming also that the body and inertia forces vanish and with account of the homogeneous deformation field we conclude that the first Piola–Kirchhoff stress field is also homogeneous. Indeed, the deformation gradient can be written as follows

$$\mathbf{F} = \lambda\mathbf{e}_1 \otimes \mathbf{e}_1 + \sqrt{\frac{J}{\lambda}}(\mathbf{e}_2 \otimes \mathbf{e}_2 + \mathbf{e}_3 \otimes \mathbf{e}_3), \tag{7.45}$$

where λ is the stretch in the direction of tension.

Then, we obtain

$$I_1 = \mathbf{F} : \mathbf{F} = \lambda^2 + \frac{2J}{\lambda}, \tag{7.46}$$

and the constitutive equations for the diagonal nonzero first Piola-Kirchhoff stress components take form

$$P_{11} = 2\frac{T}{T_0}\left(c_1 + 2c_2\left(\lambda^2 + \frac{2J}{\lambda} - 3\right) + 3c_3\left(\lambda^2 + \frac{2J}{\lambda} - 3\right)^2\right)\lambda - \frac{J}{\lambda}\Pi,$$

$$P_{22} = 2\frac{T}{T_0}\left(c_1 + 2c_2\left(\lambda^2 + \frac{2J}{\lambda} - 3\right) + 3c_3\left(\lambda^2 + \frac{2J}{\lambda} - 3\right)^2\right)\sqrt{\frac{J}{\lambda}} - \sqrt{\frac{\lambda}{J}}\Pi = P_{33}.$$

$$\tag{7.47}$$

Since the lateral stresses vanish in uniaxial tension $P_{22} = P_{33} = 0$ we can find the Lagrange parameter as follows

$$\Pi = \frac{2}{\lambda}\frac{T}{T_0}\left(c_1 + 2c_2\left(\lambda^2 + \frac{2J}{\lambda} - 3\right) + 3c_3\left(\lambda^2 + \frac{2J}{\lambda} - 3\right)^2\right). \tag{7.48}$$

Then, substituting the found parameter in the axial stress P_{11} we obtain

$$P_{11} = 2\frac{T}{T_0}\left(c_1 + 2c_2\left(\lambda^2 + \frac{2J}{\lambda} - 3\right) + 3c_3\left(\lambda^2 + \frac{2J}{\lambda} - 3\right)^2\right)\left(\lambda - \frac{J}{\lambda^2}\right),$$

$$\tag{7.49}$$

or in terms of the Cauchy stress

$$\sigma_{11} = \frac{\lambda}{J}P_{11}. \tag{7.50}$$

The constitutive equation for entropy takes the following form

$$\eta_0 = \frac{1}{T_0}\sum_{k=1}^{3} c_k\left(\lambda^2 + \frac{2J}{\lambda} - 3\right)^k - c_0\ln\frac{T}{T_0} + \frac{\partial g}{\partial T}\Pi, \tag{7.51}$$

where Π is defined by (7.48).

We choose the thermoelastic constraint in the form

$$J = f = \exp[3\gamma_0(T - T_0)], \tag{7.52}$$

where γ_0 is a constant *thermal expansion coefficient*, and, consequently, we get

$$\frac{\partial f}{\partial T} = 3\gamma_0\exp[3\gamma_0(T - T_0)]. \tag{7.53}$$

Fig. 7.2 Temperature [°K] versus stretch for constant entropy: thermoelastic inversion point at $\lambda_{inv} = 1.062$; triangles (▲) show the Joule experimental data

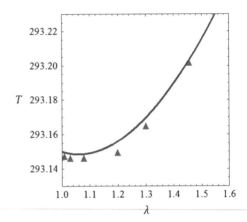

We further specify thermal constants as follows: $T_0 = 293.15°$K (20 °C), $\gamma_0 = 22.33$ 1/°K, $c_0 = 2.2$ MPa/°K.

Fixing entropy in (7.51) it is possible to find the relationship between stretches and temperature. Such a relationship is shown graphically in Fig. 7.2. The entropy magnitude was set $\eta_0 = 0.399314$ KPa/°K. to fit the room temperature to the reference state without stretches.

It is interesting to observe in Fig. 7.2 the appearance of the thermoelastic inversion point at $\lambda_{inv} = 1.062$, which was first discovered experimentally by Joule (1859). Though comparison of Joule's results with the theoretical prediction is very encouraging it should not be overstated. Indeed, to calibrate the theoretical model we used modern experimental data for material constants while Joule, most likely, used a different material.

7.7 Exercises

1. Derive (7.6).
2. Derive (7.8).
3. Draw the Cauchy stress-stretch curve for the considered example of uniaxial tension at temperatures $T = 10$ °C and $T = 30$ °C.

References

Chadwick P (1974) Thermo-mechanics of rubberlike materials. Philos Trans R Soc A 276:371–403
Chadwick P, Creasy CFM (1984) Modified entropic elasticity of rubberlike materials. J Mech Phys Solids 32:337–357
Holzapfel GA (2000) Nonlinear solid mechanics. Wiley, New York

Joule JP (1859) On some thermo-dynamic properties of solids. Philos Trans R Soc Lond A149: 91–131

Kestin J (1979) A course in thermodynamics, vol 2. CRC, Boca Raton

Maugin GA (1998) The thermomechanics of nonlinear irreversible behaviours: an introduction. World Scientific, Singapore

Muller I, Muller WH (2009) Fundamentals of thermodynamics and applications. Springer, Heidelberg

Treloar LRG (1975) The physics of rubber elasticity. OUP, Oxford

Truesdell C (1984) Rational thermodynamics. Springer, Heidelberg

Volokh KY (2015) Non-linear thermoelasticity with energy limiters. Int J Non-Linear Mech 76: 169–175

Chapter 8
Chemoelasticity

Previously, we attributed displacements, stresses, strains, temperature, entropy etc. to a material point or an infinitesimal material volume. In many cases of practical interest additional parameters reflecting the presence of the additional material/chemical *constituents*, e.g. ions, atoms, molecules etc., are required. For example, gels composed of a network of cross-linked molecules swell when a solvent migrates through it. The concentration of the solvent is changing and the gel deforms (remember diapers). When dried the gel shrinks analogously to the consolidation process in soils where the applied load enforces water to leave the solid skeleton. Soft biological tissues, like cartilage, exhibit sound alterations of the fluid phase during walking. Muscles and heart respond actively to the changing concentration of the chemical and ionic constituents.

We always assume below that the constituent *diffusion* process is slow enough to *neglect the inertia effects*.

8.1 Constituent Mass Balance

Governing equations accounting for the *chemo-mechanical* coupling should include the equations of balance and boundary conditions for the chemical/material constituents of interest. We consider *only one* additional chemical/material component of interest for the sake of simplicity. The results of Sects. 3.5 and 3.6 on Eulerian and Lagrangean forms of the master balance equations are crucial for further developments.

The integral form of the Eulerian balance law for the constituent is

$$\frac{d}{dt} \int c \, dV = \int \xi \, dV - \oint \varphi \cdot \mathbf{n} \, dA, \qquad (8.1)$$

© Springer Science+Business Media Singapore 2016
K. Volokh, *Mechanics of Soft Materials*, DOI 10.1007/978-981-10-1599-1_8

where c is the *concentration*, i.e. the number of molecules or moles, of the constituent per unit current volume; ξ is the constituent volumetric supply; and φ is the constituent flux through the current body surface with the unit outward normal $\mathbf{n}$.

Localizing this equation by getting rid of the integrals, we formulate the field balance law

$$\dot{c} + c\,\mathrm{div}\,\mathbf{v} = -\mathrm{div}\varphi + \xi, \tag{8.2}$$

in Ω and natural

$$-\varphi \cdot \mathbf{n} = \bar{\varphi}_n, \tag{8.3}$$

or essential

$$f(c) = 0 \tag{8.4}$$

boundary conditions on $\partial\Omega$, where the barred quantity is prescribed and f is a boundary condition imposed on the concentration.

The initial condition in Ω takes form

$$c(t = 0) = \bar{c}. \tag{8.5}$$

The Lagrangean or referential description of the balance law is

$$\frac{d}{dt}\int c_0 dV_0 = \int \xi_0 dV_0 - \oint \varphi_0 \cdot \mathbf{n}_0 dA_0, \tag{8.6}$$

and it can be localized in Ω_0 as

$$\frac{\partial c_0}{\partial t} = -\mathrm{Div}\varphi_0 + \xi_0. \tag{8.7}$$

Boundary conditions on $\partial\Omega_0$ read

$$-\varphi_0 \cdot \mathbf{n}_0 = \bar{\varphi}_{0n}, \\ f_0(c_0) = 0, \tag{8.8}$$

and the initial condition in Ω_0 is

$$c_0(t = 0) = \bar{c}_0. \tag{8.9}$$

We remind the reader that the Eulerian and Lagrangean quantities are related as follows (see Sect. 3.6)

$$c = J^{-1}c_0, \quad \xi = J^{-1}\xi_0, \quad \varphi = J^{-1}\mathbf{F}\varphi_0, \\ dV = J^{-1}dV_0, \quad \mathbf{n}dA = J\mathbf{F}^{-\mathrm{T}}\mathbf{n}_0 dA_0. \tag{8.10}$$

8.2 Energy Balance

Considering the energy balance we neglect the inertia effects and drop kinetic energy $\rho \mathbf{v} \cdot \mathbf{v}/2$. However, the energy necessary to diffuse the additional constituent is taken into account as follows

$$\frac{d}{dt}\int e dV = \int (\mathbf{v}\cdot\mathbf{b} + r + \mu\xi)dV + \oint (\mathbf{v}\cdot\mathbf{t} - \mathbf{q}\cdot\mathbf{n} - \mu\boldsymbol{\varphi}\cdot\mathbf{n})dA, \qquad (8.11)$$

where μ is called the *chemical potential* and it is considered as a primitive (non-reducible) object; ρ, $\mathbf{b}$, $\mathbf{t}$ are the mass density, body force, and traction respectively; e is the internal energy per unit volume; r is the rate of the heat supply per unit volume; $\mathbf{q}$ is a vector of the heat flux per unit area.

The chemical potential is a quality related to the energy that is required for the additional constituent to diffuse.

Differentiating the integral and using the stress tensor $\mathbf{t} = \sigma\mathbf{n}$, we get

$$\int (\dot{e}+e\mathrm{div}\mathbf{v})dV = \int (\mathbf{v}\cdot\mathbf{b} + r + \mu\xi)dV + \oint (\mathbf{v}\cdot\sigma\mathbf{n} - \mathbf{q}\cdot\mathbf{n} - \mu\boldsymbol{\varphi}\cdot\mathbf{n})dA. \qquad (8.12)$$

By using the divergence theorem we localize

$$\dot{e} + e\mathrm{div}\mathbf{v} = (\mathbf{v}\cdot\mathbf{b} + r + \mu\xi) + \mathrm{div}(\sigma^{\mathrm{T}}\mathbf{v} - \mathbf{q} - \mu\boldsymbol{\varphi}). \qquad (8.13)$$

By virtue of (7.5) and equilibrium condition $\mathrm{div}\sigma + \mathbf{b} = \mathbf{0}$, the local energy balance takes form

$$\dot{e} + e\mathrm{div}\mathbf{v} = \sigma : \mathbf{D} + r + \mu\xi - \mathrm{div}(\mathbf{q} + \mu\boldsymbol{\varphi}). \qquad (8.14)$$

The latter equation can be further transformed to

$$\dot{e} + e\mathrm{div}\mathbf{v} = \sigma : \mathbf{D} + r - \mathrm{div}\mathbf{q} + \mu(\xi - \mathrm{div}\boldsymbol{\varphi}) - \boldsymbol{\varphi}\cdot\mathrm{grad}\mu, \qquad (8.15)$$

or, via the balance equation (8.2), we have

$$\dot{e} + e\mathrm{div}\mathbf{v} = \sigma : \mathbf{D} + r - \mathrm{div}\mathbf{q} + \mu(\dot{c} + \mathrm{div}\mathbf{v}) - \boldsymbol{\varphi}\cdot\mathrm{grad}\mu. \qquad (8.16)$$

Alternatively, the Lagrangean description of the energy balance gives the following integral form

$$\frac{d}{dt}\int e_0 dV_0 = \int (\mathbf{v}\cdot\mathbf{b}_0 + r_0 + \mu\xi_0)dV_0 + \oint (\mathbf{v}\cdot\mathbf{t}_0 - \mathbf{q}_0\cdot\mathbf{n}_0 - \mu\boldsymbol{\varphi}_0\cdot\mathbf{n}_0)dA_0, \qquad (8.17)$$

and it can be localized as

$$\dot{e}_0 = \mathbf{P} : \dot{\mathbf{F}} + r_0 - \mathrm{Div}\mathbf{q}_0 + \mu \dot{c}_0 - \boldsymbol{\varphi}_0 \cdot \mathrm{Grad}\mu. \tag{8.18}$$

8.3   Entropy Inequality

We remind the reader that the entropy inequality (7.13) can be written as

$$\dot{\eta} + \eta \mathrm{div}\mathbf{v} \geq \frac{1}{T}(r - \mathrm{div}\mathbf{q}) + \frac{1}{T^2}\mathbf{q} \cdot \mathrm{grad}T. \tag{8.19}$$

On the other hand, we have from the generalized energy balance

$$r - \mathrm{div}\mathbf{q} = \dot{e} + e\mathrm{div}\mathbf{v} - \boldsymbol{\sigma} : \mathbf{D} - \mu(\dot{c} + \mathrm{div}\mathbf{v}) + \boldsymbol{\varphi} \cdot \mathrm{grad}\mu. \tag{8.20}$$

Substituting the latter equation in (8.19) we get the generalized Clausius–Duhem inequality

$$\dot{\eta} + \eta \mathrm{div}\mathbf{v} \geq \frac{1}{T}(\dot{e} + e\mathrm{div}\mathbf{v} - \boldsymbol{\sigma} : \mathbf{D} - \mu(\dot{c} + \mathrm{div}\mathbf{v}) + \boldsymbol{\varphi} \cdot \mathrm{grad}\mu) + \frac{1}{T^2}\mathbf{q} \cdot \mathrm{grad}T. \tag{8.21}$$

Its Lagrangean form is

$$\dot{\eta}_0 \geq \frac{1}{T}(\dot{e}_0 - \mathbf{P} : \dot{\mathbf{F}} - \mu \dot{c}_0 + \boldsymbol{\varphi}_0 \cdot \mathrm{Grad}\mu) + \frac{1}{T^2}\mathbf{q}_0 \cdot \mathrm{Grad}T. \tag{8.22}$$

The Eulerian and Lagrangean forms of the reduced generalized Clausius–Planck inequality are

$$\dot{\eta} + \eta \mathrm{div}\mathbf{v} \geq \frac{1}{T}(\dot{e} + e\mathrm{div}\mathbf{v} - \boldsymbol{\sigma} : \mathbf{D} - \mu(\dot{c} + \mathrm{div}\mathbf{v}) + \boldsymbol{\varphi} \cdot \mathrm{grad}\mu), \tag{8.23}$$

and

$$\dot{\eta}_0 \geq \frac{1}{T}(\dot{e}_0 - \mathbf{P} : \dot{\mathbf{F}} - \mu \dot{c}_0 + \boldsymbol{\varphi}_0 \cdot \mathrm{Grad}\mu). \tag{8.24}$$

Equivalently, it is possible to require the non-negative generalized internal dissipation

$$D_{\mathrm{int}} = T\dot{\eta}_0 - \dot{e}_0 + \mathbf{P} : \dot{\mathbf{F}} + \mu \dot{c}_0 - \boldsymbol{\varphi}_0 \cdot \mathrm{Grad}\mu \geq 0. \tag{8.25}$$

8.4 Helmholtz's Free Energy Function

Let us introduce the Helmholtz free energy function in the form

$$\psi(\mathbf{F}, c_0, T) = e_0 - T\eta_0. \qquad (8.26)$$

Differentiating this equation with respect to time we obtain

$$\frac{\partial \psi}{\partial \mathbf{F}} : \dot{\mathbf{F}} + \frac{\partial \psi}{\partial c_0} \dot{c}_0 + \frac{\partial \psi}{\partial T} \dot{T} = \dot{e}_0 - \dot{T}\eta_0 - T\dot{\eta}_0. \qquad (8.27)$$

Substitution of (8.27) in the dissipation inequality (8.25) yields

$$D_{\text{int}} = \left(\mathbf{P} - \frac{\partial \psi}{\partial \mathbf{F}}\right) : \dot{\mathbf{F}} - \left(\eta_0 + \frac{\partial \psi}{\partial T}\right) \dot{T} + \left(\mu - \frac{\partial \psi}{\partial c_0}\right) \dot{c}_0 - \boldsymbol{\varphi}_0 \cdot \text{Grad}\mu \geq 0. \qquad (8.28)$$

Based on this inequality and following the Coleman-Noll procedure we introduce the constitutive laws in the form

$$\mathbf{P} = \frac{\partial \psi}{\partial \mathbf{F}},$$
$$\eta_0 = -\frac{\partial \psi}{\partial T}, \qquad (8.29)$$
$$\mu = \frac{\partial \psi}{\partial c_0}.$$

With account of these constitutive laws the dissipation inequality reduces to

$$- \boldsymbol{\varphi}_0 \cdot \text{Grad}\mu \geq 0. \qquad (8.30)$$

Substitution of (8.27) in (8.18) yields

$$T\dot{\eta}_0 = \left(\mathbf{P} - \frac{\partial \psi}{\partial \mathbf{F}}\right) : \dot{\mathbf{F}} - \left(\eta_0 + \frac{\partial \psi}{\partial T}\right) \dot{T} + \left(\mu \dot{c}_0 - \frac{\partial \psi}{\partial c_0} \dot{c}_0\right) + r_0 - \text{Div}\mathbf{q}_0 - \boldsymbol{\varphi}_0 \cdot \text{Grad}\mu, \qquad (8.31)$$

and by virtue of (8.29) we have

$$T\dot{\eta}_0 = r_0 - \text{Div}\mathbf{q}_0 - \boldsymbol{\varphi}_0 \cdot \text{Grad}\mu. \qquad (8.32)$$

We note that the entropy increment can be calculated from the constitutive law as follows

$$\dot{\eta}_0 = -\frac{\partial^2 \psi}{\partial T \partial \mathbf{F}} : \dot{\mathbf{F}} - \frac{\partial^2 \psi}{\partial T^2} \dot{T} - \frac{\partial^2 \psi}{\partial T \partial c_0} \dot{c}_0. \qquad (8.33)$$

Substituting (8.33) in (8.32) we finally get the thermo-chemo-mechanical energy balance in the form

$$-T\frac{\partial^2\psi}{\partial T^2}\dot{T} = r_0 - \text{Div}\mathbf{q}_0 + T\frac{\partial^2\psi}{\partial T\partial\mathbf{F}} : \dot{\mathbf{F}} + T\frac{\partial^2\psi}{\partial T\partial c_0}\dot{c}_0 - \boldsymbol{\varphi}_0 \cdot \text{Grad}\mu. \quad (8.34)$$

8.5   Fick's Law of Diffusion

The *Fick law* of diffusion is analogous to the Fourier law of heat conduction and it can be written in the following form

$$\boldsymbol{\varphi} = -\mathbf{M}\text{grad}\mu, \quad (8.35)$$

where $\mathbf{M}$ is called the *mobility tensor.*

Transforming the Fick law to the Lagrangean description we get

$$\boldsymbol{\varphi}_0 = -\mathbf{M}_0\text{Grad}\mu, \quad (8.36)$$

where $\mathbf{M}_0$ is the referential mobility tensor.

$$\mathbf{M}_0 = J\mathbf{F}^{-1}\mathbf{M}\mathbf{F}^{-\text{T}}. \quad (8.37)$$

The Fick law is essentially a constitutive law and it should obey the dissipation inequality (8.30) as follows

$$\text{Grad}\mu \cdot \mathbf{M}_0\text{Grad}\mu \geq 0, \quad (8.38)$$

which implies that the mobility tensor $\mathbf{M}_0$ must be positive semi-definite.

8.6   Chemoelastic Incompressibility

Generalizing the notion of elastic or thermoelastic incompressibility we introduce the *chemo-thermo-elastic incompressibility* constraint in the following form

$$J = f(T, c_0),$$
$$f(T_0, \bar{c}_0) = 1, \quad (8.39)$$

where $\bar{c}_0$ is the given concentration of the constituent in the referential state Ω_0.

Thus, material is incompressible in the reference state and all volumetric changes are produced by the thermal and chemical alterations.

The time increment of this constraint takes form

$$\frac{\partial f}{\partial T}\dot{T} + \frac{\partial f}{\partial c_0}\dot{c}_0 - J\mathbf{F}^{-\mathrm{T}} : \dot{\mathbf{F}} = 0. \tag{8.40}$$

Here multipliers $\partial f/\partial T$, $\partial f/\partial c_0$, and $-J\mathbf{F}^{-\mathrm{T}}$ represent workless entropy, chemical potential, and stress accordingly. All these workless quantities can be scaled by arbitrary parameter Π. Thus, we modify constitutive equation (8.29) as follows

$$
\begin{aligned}
\mathbf{P} &= \frac{\partial \psi}{\partial \mathbf{F}} - J\mathbf{F}^{-\mathrm{T}}\Pi, \\
\eta_0 &= -\frac{\partial \psi}{\partial T} + \frac{\partial f}{\partial T}\Pi, \\
\mu &= \frac{\partial \psi}{\partial c_0} + \frac{\partial f}{\partial c_0}\Pi.
\end{aligned}
\tag{8.41}
$$

8.7 Diffusion Through Polymer Membrane

Based on the described theoretical framework and neglecting thermal effects, we examine the problem of diffusion of a liquid (toluene) through a polymer membrane—Fig. 8.1. This problem was considered experimentally by Paul and Ebra-Lima (1970).

A thin polymer layer is placed on a permeable porous plate and the liquid diffuses through the membrane under pressure $p_2 > p_1$. We assume that the body force and the liquid source are zero and the process is steady: $\dot{c}_0 = 0$. Under these assumptions the balance equations reduce to

$$
\begin{aligned}
\mathrm{Div}\boldsymbol{\varphi}_0 &= 0, \\
\mathrm{Div}\mathbf{P} &= \mathbf{0}.
\end{aligned}
\tag{8.42}
$$

It is further assumed that the ground material (polymer) in the reference state is incompressible and the volume of the material is altering only due to the supply of

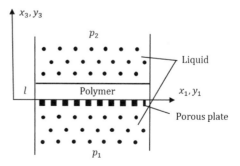

Fig. 8.1 Diffusion through polymer membrane

new species (molecules of the liquid). This assumption can be formalized by using
the following chemoelastic incompressibility constraint

$$J = 1 + \nu c_0, \tag{8.43}$$

where ν is the volume of one molecule of the liquid.

With account of the chemoelastic incompressibility constraint we have constitutive equations in the form

$$\mathbf{P} = \frac{\partial \psi}{\partial \mathbf{F}} - J \Pi \mathbf{F}^{-T},$$
$$\mu = \frac{\partial \psi}{\partial c_0} + \nu \Pi. \tag{8.44}$$

Since the thickness of the membrane is small as compared to the characteristic
lengths of the device, we can consider the field variations in the lateral directions
only. Specifically, we assume the following deformation and concentration gradients
respectively

$$\mathbf{F} = \mathbf{e}_1 \otimes \mathbf{e}_1 + \mathbf{e}_2 \otimes \mathbf{e}_2 + \lambda(x_3)\mathbf{e}_3 \otimes \mathbf{e}_3,$$
$$\boldsymbol{\varphi}_0 = \varphi_{03}(x_3)\mathbf{e}_3. \tag{8.45}$$

With account of these assumptions we get the following non-trivial stress and flux
components

$$P_{11} = \frac{\partial \psi}{\partial F_{11}} - \lambda \Pi = P_{22},$$
$$P_{33} = \frac{\partial \psi}{\partial F_{33}} - \Pi, \tag{8.46}$$
$$\varphi_{03} = -M_{033}\frac{\partial \mu}{\partial x_3}.$$

We note that the traction and placement boundary conditions take the following
forms on the upper and lower surfaces of the membrane accordingly

$$P_{33}(L) = -p_2,$$
$$y_3(0) = 0. \tag{8.47}$$

Since the stress tensor is divergence-free and, consequently, $P_{33} = $ constant, we
can obtain the Lagrange parameter Π from the first boundary condition

$$\Pi = \frac{\partial \psi}{\partial F_{33}} + p_2. \tag{8.48}$$

Substituting the Lagrange parameter in the constitutive law for the chemical potential we get

$$\mu = \frac{\partial \psi}{\partial c_0} + \nu \frac{\partial \psi}{\partial F_{33}} + \nu p_2. \tag{8.49}$$

We also notice that due to the chemoelastic incompressibility condition the liquid concentration is related to the stretch as follows

$$\nu c_0 = \lambda - 1. \tag{8.50}$$

We define the mobility tensor and the Helmholtz free energy function as follows

$$\mathbf{M}_0 = (\alpha c_0 \nu)^{\beta-1} \frac{c_0 D}{kT} \mathbf{C}^{-1}, \tag{8.51}$$

$$\psi = \frac{1}{2} N k T \{ \mathbf{F} : \mathbf{F} - 3 - 2 \log J \} - \frac{kT}{\nu} \left\{ \nu c_0 \log \left(1 + \frac{1}{c_0 \nu} \right) + \frac{\chi}{1 + c_0 \nu} \right\}, \tag{8.52}$$

where the first term on the right-hand side of the Helmholtz free energy is the elastic *strain energy* and the second one is the *energy of mixing*; α and β are dimensionless material constants; D is the diffusion coefficient for the solvent molecules; k is the Boltzmann constant; T is a (constant) absolute temperature; N is the number of polymer chains in the gel divided by the reference volume; and χ is a dimensionless parameter.

With account of the mobility tensor and the Helmholtz free energy function we obtain

$$\frac{\partial \psi}{\partial F_{33}} = N k T (\lambda - \lambda^{-1}), \tag{8.53}$$

and

$$\varphi_{03} = -(\alpha c_0 \nu)^{\beta-1} \frac{c_0 D}{kT} \lambda^{-2} \frac{\partial \mu}{\partial x_3}, \tag{8.54}$$

and

$$\mu = kT \left(\log \frac{\lambda - 1}{\lambda} + \frac{1}{\lambda} + \frac{\chi}{\lambda^2} \right) + N \nu k T (\lambda - \lambda^{-1}) + \nu p_2. \tag{8.55}$$

Substituting (8.54) and (8.55) in (8.42)$_1$ we get a second order ordinary differential equation of the chemical balance in term of stretches. To solve it we need to impose two boundary conditions on the chemical potential

$$\begin{aligned} \mu(\lambda_1) &= p_1 \nu, \\ \mu(\lambda_2) &= p_2 \nu, \end{aligned} \tag{8.56}$$

where $\lambda_1 = \lambda(0)$ and $\lambda_2 = \lambda(L)$.

Fig. 8.2 Theory and experiment (▲) for flux $[\frac{cm^3}{cm^2 day}]$ versus pressure [psi] in the diffusion problem

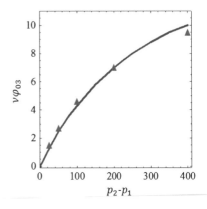

Table 8.1 Toluene-polymer data

Parameter	Magnitude
k	1.38×10^{-23} Nm/K
T	$303°$ K
p_1	10^5 N/m^2
ν	17.7×10^{-29} m^3
D	2.36×10^{-10} m^2/s
N	6.36×10^{25} 1/m^3
L	2.65×10^{-4} m
χ	0.425
α	5.7
β	3

Based on the numerical solution of the two-point boundary-value problem it is possible to compute the increase of the flux through the membrane with the increase of the pressure—see Fig. 8.2 for the toluene-polymer data shown in Table 8.1.

Remarkably, the flux increase is not proportional to the pressure increase!

8.8 Exercises

1. Derive (8.54) and (8.55).
2. Write an explanation of the physical meaning of the *chemical potential* based on a literature review.

References

Baek S, Srinivasa AR (2004) Diffusion of a fluid through an elastic solid undergoing large deformation. Int J Non-Linear Mech 39:201–218

Flory PJ, Rehner J (1944) Effect of the deformation on the swelling capacity of the rubber. J Chem Phys 12:413–414

Gibbs JW (1878) The scientific papers of J. Willard Gibbs, vol 1. On the equilibrium of heterogeneous substances. Dover, New York, 1961

Gurtin ME, Fried E, Anand L (2010) The mechanics and thermodynamics of continua. Cambridge University Press, Cambridge

Hong W, Zhao X, Zhou J, Suo Z (2008) A theory of coupled diffusion and large deformation in polymeric gels. J Mech Phys Solids 56:1779–1793

Paul DR, Ebra-Lima OO (1970) Pressure-induced diffusion of organic liquids through highly swollen polymer membranes. J Appl Polym Sci 14:2201–2224

Treloar LRG (1975) The physics of rubber elasticity. Oxford University Press, Oxford

Volokh KY (2012) On diffusion through soft filter. J Appl Mech 79:064503

Chapter 9
Electroelasticity

Soft polymer materials are *dielectric*, i.e. they do not conduct the electric current. However, *electroactive polymers* (EAP) can deform in response to electric fields. This property is increasingly used in actuators or artificial muscles that have a great potential of practical applications. In this chapter, we consider the basic quasi-static electroelasticity of soft materials at large strains.

9.1 Electrostatics

Electron presents the smallest negative charge of 1.6×10^{-19} C (Coulomb). All other charges, both positive and negative, are multipliers of the electron charge. The charges can be free, leading to the electric current, or bound as in the case of electroactive dielectrics. Since the number of charges in the material volumes that we consider is large, we will always mean the continuum average in the subsequent considerations.

Charges create electric fields that produce forces on other charges. For example, the force on charge Q is

$$\mathbf{f} = Q\mathbf{e}, \tag{9.1}$$

where $\mathbf{e}$ is the *electric field*.

According to the experimentally validated Coulomb's law the force between charges Q and Q' placed at points $\mathbf{y}$ and $\mathbf{y}'$ accordingly is inversely proportional to the squared distance between the charges

$$\mathbf{f} = Q\frac{Q'}{4\pi\varepsilon_0}\frac{\mathbf{y} - \mathbf{y}'}{|\mathbf{y} - \mathbf{y}'|^3}, \tag{9.2}$$

© Springer Science+Business Media Singapore 2016
K. Volokh, *Mechanics of Soft Materials*, DOI 10.1007/978-981-10-1599-1_9

where $\varepsilon_0 = 8.854 \times 10^{-12}$ F/m (Farad/meter) is called the *electric permittivity* of space.

Comparing two previous formulas we conclude that the electric field defined by the Coulomb law is

$$\mathbf{e} = \frac{Q'}{4\pi\varepsilon_0} \frac{\mathbf{y} - \mathbf{y}'}{|\mathbf{y} - \mathbf{y}'|^3}. \tag{9.3}$$

For multiple charges, we can rewrite the electric field by using superposition and smearing over the space with the charge density q

$$\mathbf{e}(\mathbf{y}) = \frac{1}{4\pi\varepsilon_0} \int q(\mathbf{y}') \frac{\mathbf{y} - \mathbf{y}'}{|\mathbf{y} - \mathbf{y}'|^3} dV'. \tag{9.4}$$

Based on the latter equation we can obtain (without proof) the *Gauss law* for a space volume V enclosed with a surface A with the outward unit normal $\mathbf{n}$

$$\oint \varepsilon_0 \mathbf{e} \cdot \mathbf{n} dA = \int q dV. \tag{9.5}$$

The Gauss law was derived for vacuum in the absence of matter. In the presence of matter, the bound charges can be slightly displaced with respect to each other when the electric potential is applied—Fig. 9.1.

Such relative displacement causes *polarization*. To characterize the phenomenon it is possible to introduce the *polarization vector* $\mathbf{p}$, which gathers and averages individual *dipole moments* over a unit volume. The polarization vector changes the charge on the right hand side of the Gauss law

$$\oint \varepsilon_0 \mathbf{e} \cdot \mathbf{n} dA = \int q dV - \oint \mathbf{p} \cdot \mathbf{n} dA. \tag{9.6}$$

It is convenient to introduce the *electric displacement vector*

$$\mathbf{d} = \varepsilon_0 \mathbf{e} + \mathbf{p} \tag{9.7}$$

Fig. 9.1 Polarization: more molecular charge leaves unit volume than enters it

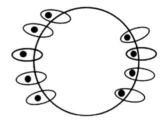

and rewrite the Gauss law in the form

$$\oint \mathbf{d} \cdot \mathbf{n} dA = \int q dV. \tag{9.8}$$

This equation is valid for any volume and, consequently, we can localize it

$$\operatorname{div}\mathbf{d} = q. \tag{9.9}$$

Two last formulas represent the integral and differential forms of the *first equation of electrostatics*.

To derive the second equation of electrostatics we note that

$$\frac{\mathbf{y} - \mathbf{y}'}{|\mathbf{y} - \mathbf{y}'|^3} = -\operatorname{grad}(|\mathbf{y} - \mathbf{y}'|^{-1}). \tag{9.10}$$

Substituting this identity in (9.4) we obtain

$$\mathbf{e}(\mathbf{y}) = -\operatorname{grad}\varphi, \tag{9.11}$$

where

$$\varphi(\mathbf{y}) = -\frac{1}{4\pi\varepsilon_0} \int \frac{q(\mathbf{y}')}{|\mathbf{y} - \mathbf{y}'|} dV' \tag{9.12}$$

is called the *electric potential* or *voltage*.

To clarify the physical meaning of the electric potential we consider the work that should be done against the electric field in order to move charge Q from point $\mathbf{y}_1$ to point $\mathbf{y}_2$

$$-\int_{\mathbf{y}_1}^{\mathbf{y}_2} Q\mathbf{e} \cdot d\mathbf{y} = \int_{\mathbf{y}_1}^{\mathbf{y}_2} Q(\operatorname{grad}\varphi) \cdot d\mathbf{y} = Q\varphi(\mathbf{y}_2) - Q\varphi(\mathbf{y}_1). \tag{9.13}$$

Thus, the work is equal to the difference in the electrical potentials at points $\mathbf{y}_1$ and $\mathbf{y}_2$ times charge Q. Since the integral does not depend on the integration path we have for closed curve l

$$\oint \mathbf{e} \cdot d\mathbf{y} = 0. \tag{9.14}$$

By building any surface A on the curve l and using the Stokes formula (1.64) we can rewrite the latter equation in the form

$$\oint \mathbf{e} \cdot d\mathbf{y} = \int (\operatorname{curl}\mathbf{e}) \cdot \mathbf{n} dA = 0. \tag{9.15}$$

Since the surface can be chosen arbitrarily, we can localize the integral equation as follows

$$\text{curle} = \mathbf{0}. \tag{9.16}$$

We note that the electric field derived from the electric potential always obeys the latter equation which transforms to identity. Two last formulas present the integral and differential forms of the *second equation of electrostatics*.

Let us derive the boundary conditions on a surface separating two materials with different electric fields and displacements—Fig. 9.2.

Firstly, we consider a small cylinder with the base area $\triangle A$ and height $h \to 0$. In this case, the left- and right- hand sides of (9.8) take the following forms accordingly

$$\oint \mathbf{d} \cdot \mathbf{n} dA = \mathbf{d}^{(2)} \cdot \mathbf{n}\triangle A + \mathbf{d}^{(1)} \cdot (-\mathbf{n})\triangle A = (\mathbf{d}^{(2)} - \mathbf{d}^{(1)}) \cdot \mathbf{n}\triangle A, \tag{9.17}$$

and

$$\int q dV = q_A \triangle A, \tag{9.18}$$

where q_A is a charge on the boundary surface.

Substituting these equations back in (9.8) we can write the following boundary condition

$$(\mathbf{d}^{(2)} - \mathbf{d}^{(1)}) \cdot \mathbf{n} = q_A. \tag{9.19}$$

Secondly, we consider a closed path, l, whose long arm directions are defined by the cross product of the surface tangent $\mathbf{s}$ and normal $\mathbf{n}$ vectors. In this case (9.14) takes the following form

$$\oint \mathbf{e} \cdot d\mathbf{y} = \mathbf{e}^{(1)} \cdot (\mathbf{n} \times \mathbf{s})\triangle l - \mathbf{e}^{(2)} \cdot (\mathbf{n} \times \mathbf{s})\triangle l = \triangle l \mathbf{s} \cdot (\mathbf{e}^{(1)} \times \mathbf{n} - \mathbf{e}^{(2)} \times \mathbf{n}) = 0. \tag{9.20}$$

Fig. 9.2 Boundary surface

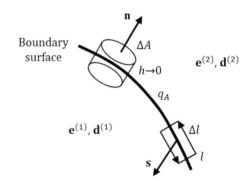

Since this equation is correct for any tangent vector $\mathbf{s}$ we obtain the second boundary condition

$$(\mathbf{e}^{(1)} - \mathbf{e}^{(2)}) \times \mathbf{n} = \mathbf{0}. \tag{9.21}$$

Finally, we note that the polarization vector should be defined as a function of the electric field or, in other words, the constitutive equation should be written in the form $\mathbf{p}(\mathbf{e})$.

The simplest form of the constitutive equation in the case of isotropic media is the linear proportionality between polarization and electric field

$$\mathbf{p} = \omega \varepsilon_0 \mathbf{e}, \tag{9.22}$$

where ω is called the *electric susceptibility* of the medium.

With account of this constitutive law the electric displacement takes form

$$\mathbf{d} = \varepsilon \mathbf{e}, \tag{9.23}$$

where

$$\varepsilon = \varepsilon_0 \varepsilon_r \tag{9.24}$$

is called the *electric permittivity* and ratio and

$$\varepsilon_r = 1 + \omega \tag{9.25}$$

is called the *dielectric constant*.

9.2 Weak Electro-Mechanical Coupling

Electric fields create body forces and couples on dipoles embedded in materials. Though there are a number of theories defining the constitutive equation for the electric body force $\mathbf{b}_e$ and couple $\mathbf{k}$, all of them reduce to the same form in the case of electrostatics and zero distributed body charge, $q = 0$,

$$\mathbf{b}_e = (\text{grad}\,\mathbf{e})\mathbf{p}, \\ \mathbf{k} = \mathbf{p} \times \mathbf{e}. \tag{9.26}$$

Equations of the angular momentum balance (Sect. 3.4) should be modified to include the *body couple* due to the electric field

$$\frac{d}{dt} \int \rho \mathbf{r} \times \mathbf{v} dV = \int (\mathbf{r} \times (\mathbf{b} + \mathbf{b}_e) + \mathbf{k}) dV + \oint \mathbf{r} \times \mathbf{t} dA. \tag{9.27}$$

Localizing this equation as it was done in Sect. 3.4 we obtain

$$\varepsilon_{ijk}\sigma_{kj} + k_i = 0. \tag{9.28}$$

This equation means that, generally, the Cauchy stress is not symmetric anymore in the presence of the electric field: $\sigma \neq \sigma^{\mathrm{T}}$.

We call electromechanical coupling *weak* if the constitutive equations for the mechanical and electrical interactions are separated

$$\sigma(\mathbf{F}), \quad \mathbf{p}(\mathbf{e}).$$

9.3   Strong Electro-Mechanical Coupling

We call electromechanical coupling *strong* if the constitutive equations for the mechanical and electrical interactions are coupled in the following way, for example,

$$\sigma(\mathbf{F}, \mathbf{e}), \quad \mathbf{p}(\mathbf{F}, \mathbf{e}).$$

In the case of the strong coupling it is reasonable to formulate a unified electro-mechanical constitutive theory by introducing a free energy function $\psi(\mathbf{F}, \mathbf{e})$ analogously to thermo- and chemo- elastic interactions considered in the previous chapters. Such unification is carefully studied in the monograph by Dorfmann and Ogden (2014), for example. Unfortunately, materials exhibiting the strong electroelastic coupling are not well-known yet. Available experiments with the reliable results show that the effect of the strong electro-mechanical coupling is absent or negligible. Nevertheless, we examine a possibility of the strong coupling in Sect. 9.5.

9.4   Maxwell Stress and Total Stress

Following Maxwell's idea for magnetism, it is possible to represent the electric body force as a divergence of the *Maxwell stress* tensor σ_{M}

$$\mathbf{b}_{\mathrm{e}} = (\mathrm{grad}\,\mathbf{e})\mathbf{p} = \mathrm{div}\,\sigma_{\mathrm{M}}. \tag{9.29}$$

Such representation is not unique and the following popular form can be used, for example,

$$\sigma_{\mathrm{M}} = \mathbf{e} \otimes \mathbf{d} - \frac{\varepsilon_0}{2}(\mathbf{e} \cdot \mathbf{e})\mathbf{1} = \mathbf{e} \otimes \mathbf{p} + \varepsilon_0 \mathbf{e} \otimes \mathbf{e} - \frac{\varepsilon_0}{2}(\mathbf{e} \cdot \mathbf{e})\mathbf{1}. \tag{9.30}$$

Combining elastic and Maxwell stresses it is possible to introduce the *total stress*

$$\tau = \sigma + \sigma_{\mathrm{M}}, \tag{9.31}$$

which obeys the equilibrium equation without body forces

$$\text{div}\boldsymbol{\tau} = \mathbf{0}. \tag{9.32}$$

In terms of the total stress, the angular momentum balance takes the simplest form

$$\varepsilon_{ijk}\tau_{kj} = 0, \tag{9.33}$$

which shows that the total stress is symmetric, contrary to the Cauchy stress

$$\boldsymbol{\tau} = \boldsymbol{\tau}^{\mathrm{T}}. \tag{9.34}$$

We note, however, that the body couple is zero in the case of the constitutive equation (9.22)

$$\mathbf{k} = \mathbf{p} \times \mathbf{e} = \omega\varepsilon_0\mathbf{e} \times \mathbf{e} = \mathbf{0}, \tag{9.35}$$

and in the latter case the Cauchy stress is symmetric too.

9.5   Dielectric Actuator

We consider a typical deformation of a dielectric actuator shown in Fig. 9.3. Thin elastomer plate is covered with two very thin electrodes. When electrodes are activated by the applied voltage the elastomer deforms and gets thinner. We analyze this deformation below.

We assume the linear constitutive law for polarization $\mathbf{p} = \omega\varepsilon_0\mathbf{e}$. In this case, the boundary-value problem of electrostatics is presented by the following field equations

$$\begin{aligned} \text{div}\mathbf{e} &= 0, \\ \text{curl}\mathbf{e} &= \mathbf{0}, \end{aligned} \tag{9.36}$$

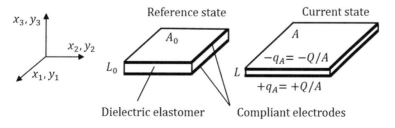

Fig. 9.3 Dielectric actuator

with boundary conditions

$$\varepsilon_0(\mathbf{e}^\star - \varepsilon_r\mathbf{e}) \cdot \mathbf{n} = q_A,$$
$$(\mathbf{e} - \mathbf{e}^\star) \times \mathbf{n} = \mathbf{0}, \tag{9.37}$$

where $\mathbf{e}$ and $\mathbf{e}^\star$ are electric fields inside and outside the plate respectively.

Momenta balance equations are

$$\mathrm{div}\boldsymbol{\tau} = \mathbf{0},$$
$$\boldsymbol{\tau} = \boldsymbol{\tau}^\mathrm{T}, \tag{9.38}$$

with the corresponding traction boundary conditions

$$(\boldsymbol{\tau} - \boldsymbol{\tau}^\star)\mathbf{n} = \mathbf{0}, \tag{9.39}$$

where $\boldsymbol{\tau}$ and $\boldsymbol{\tau}^\star$ are stress fields inside and outside the plate accordingly.

Constitutive law for the total stress of isotropic incompressible hyperelastic material can be written as

$$\boldsymbol{\tau} = -\Pi\mathbf{1} + 2(\psi_1 + I_1\psi_2)\mathbf{B} - 2\psi_2\mathbf{B}^2 + \varepsilon_0\varepsilon_r\mathbf{e} \otimes \mathbf{e} - \frac{\varepsilon_0}{2}(\mathbf{e} \cdot \mathbf{e})\mathbf{1}, \tag{9.40}$$

where Π is the Lagrange parameter; $\psi_a \equiv \partial\psi/\partial I_a$ with I_a the principal invariants of $\mathbf{B} = \mathbf{F}\mathbf{F}^\mathrm{T}$.

Constitutive law for the total stress outside the dielectric is

$$\boldsymbol{\tau}^\star = \varepsilon_0\mathbf{e}^\star \otimes \mathbf{e}^\star - \frac{\varepsilon_0}{2}(\mathbf{e}^\star \cdot \mathbf{e}^\star)\mathbf{1}. \tag{9.41}$$

The electric fields created by the electrodes outside the plate mutually compensate each other and, consequently we assume a uniform electric field inside the plate only

$$\mathbf{e} = E\mathbf{e}_3,$$
$$\mathbf{e}^\star = \mathbf{0}. \tag{9.42}$$

We also assume that stretching is uniform and the deformation gradient is diagonal

$$\mathbf{F} = \lambda^{-1/2}(\mathbf{e}_1 \otimes \mathbf{e}_1 + \mathbf{e}_2 \otimes \mathbf{e}_2) + \lambda\mathbf{e}_3 \otimes \mathbf{e}_3, \tag{9.43}$$

where the lateral stretch is

$$\lambda = \frac{L}{L_0}. \tag{9.44}$$

We note that material is incompressible, $J = 1$, and, consequently, we have

$$AL = A_0L_0,$$
$$A\lambda = A_0. \tag{9.45}$$

The electric field E can be readily obtained from the boundary conditions

$$E = \frac{q_A}{\varepsilon_0 \varepsilon_r} = \frac{Q}{\varepsilon_0 \varepsilon_r A} = \frac{Q\lambda}{\varepsilon_0 \varepsilon_r A_0}. \tag{9.46}$$

Consequently, the nonzero stresses take form

$$\tau_{11} = -\Pi + 2(\psi_1 + I_1\psi_2)\lambda^{-1} - 2\psi_2\lambda^{-2} - \frac{\varepsilon_0}{2}\left(\frac{Q\lambda}{\varepsilon_0\varepsilon_r A_0}\right)^2 = \tau_{22},$$

$$\tau_{33} = -\Pi + 2(\psi_1 + I_1\psi_2)\lambda^{-1} - 2\psi_2\lambda^{-2} + \varepsilon_0^2\varepsilon_r\left(\frac{Q\lambda}{\varepsilon_0\varepsilon_r A_0}\right)^2 - \frac{\varepsilon_0}{2}\left(\frac{Q\lambda}{\varepsilon_0\varepsilon_r A_0}\right)^2. \tag{9.47}$$

Assuming the stress-free deformation, $\tau_{11} = \tau_{22} = \tau_{33} = 0$ and excluding parameter Π from the last two equations we get

$$\frac{2\varepsilon_r}{\lambda^2}(\psi_1 + I_1\psi_2)(\lambda^{-1} - \lambda^2) - 2\psi_2(\lambda^{-2} - \lambda^4) = \frac{Q^2}{A_0^2}. \tag{9.48}$$

This equation allows us to calculate the lateral stretch λ and voltage $\Phi = EL$ for the given charge Q.

We consider three elastic strain energies for the acrylic elastomer VHB 4910. The first one is the neo-Hookean strain energy

$$\psi = c_1(I_1 - 3), \tag{9.49}$$

where $c_1 = 0.08$ MPa.

The second one is the Yeoh strain energy

$$\psi = c_1(I_1 - 3) + c_2(I_1 - 3)^2 + c_3(I_1 - 3)^3, \tag{9.50}$$

where $c_1 = 0.08$ MPa, $c_1 = 0.08$ MPa, $c_1 = 0.08$ MPa (Wissler and Mazza 2007).

The third one is the *Arruda–Boyce* strain energy in the form of the truncated series

$$\psi = c_1\left\{\frac{1}{2}(I_1 - 3) + \frac{1}{20N}(I_1^2 - 9) + \frac{11}{1050N^2}(I_1^3 - 27)\right.$$
$$\left. + \frac{19}{7000N^3}(I_1^4 - 81) + \frac{519}{673750N^4}(I_1^5 - 243)\right\}, \tag{9.51}$$

where $c_1 = 0.0686$ MPa, $N = 124.88$ (Wissler and Mazza 2007).

In addition to the purely elastic strain energies we consider a strong electromechanical coupling via the dependence of the polarization on both the deformation and electric fields, $\mathbf{p}(\mathbf{F}, \mathbf{e})$. Specifically, we assume that the dielectric constant is a function of the invariants of the Cauchy–Green tensor

$$\varepsilon_r(I_1, I_2).$$

We further make the simplest assumption that the dielectric parameter depends linearly on the first invariant only

$$\varepsilon_r = \alpha_0 - \alpha_1(I_1 - 3), \tag{9.52}$$

where $\varepsilon_r = \alpha_0$ for $I_1 = 3$ in the absence of deformation.

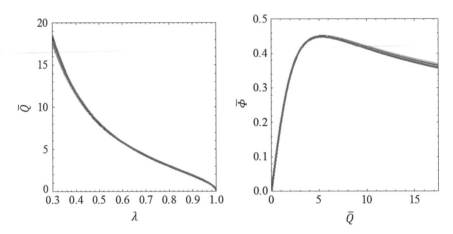

Fig. 9.4 Dimensionless charge $\bar{Q} = Q/(A_0\sqrt{c_1})$ versus lateral stretch λ and dimensionless voltage $\bar{\Phi} = \Phi\varepsilon_0/(L_0\sqrt{c_1})$ for neo-Hookean model: *yellow line* for $\alpha_1 = 0.1$; *blue line* for $\alpha_1 = 0.049$; *green line* for $\alpha_1 = 0.02$; *red line* for $\alpha_1 = 0.0$

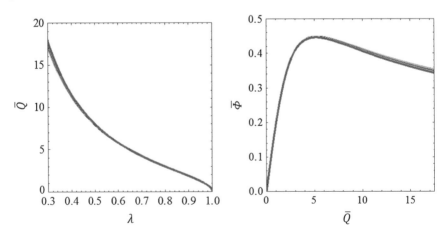

Fig. 9.5 Dimensionless charge $\bar{Q} = Q/(A_0\sqrt{c_1})$ versus lateral stretch λ and dimensionless voltage $\bar{\Phi} = \Phi\varepsilon_0/(L_0\sqrt{c_1})$ for Yeoh model: *yellow line* for $\alpha_1 = 0.1$; *blue line* for $\alpha_1 = 0.049$; *green line* for $\alpha_1 = 0.02$; *red line* for $\alpha_1 = 0.0$

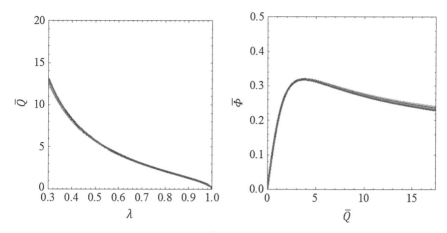

Fig. 9.6 Dimensionless charge $\bar{Q} = Q/(A_0\sqrt{c_1})$ versus lateral stretch λ and dimensionless voltage $\bar{\Phi} = \Phi\varepsilon_0/(L_0\sqrt{c_1})$ for Arruda–Boyce model: *yellow line* for $\alpha_1 = 0.1$; *blue line* for $\alpha_1 = 0.049$; *green line* for $\alpha_1 = 0.02$; *red line* for $\alpha_1 = 0.0$

In simulations we assume $\alpha_0 = 4.7$ and four possible cases for $\alpha_1 = 0.1$; 0.049; 0.02; 0.0. In the latter case of $\alpha_1 = 0.0$ the electro-mechanical coupling is weak.

Figures 9.4, 9.5, and 9.6 present the results of calculations for the chosen material models.

The results show that the permittivity alterations (strong electro-mechanical coupling) of the dielectric actuator during deformation practically do not affect the process of *electrostriction*!

9.6 Exercises

1. Prove (9.10).
2. Check whether (9.30) obeys (9.29).
3. Derive (9.33).

References

Dorfmann A, Ogden RW (2014) Nonlinear theory of electroelastic and magnetoelastic interactions. Springer, New York

Eringen AC, Maugin GA (1989) Electrodynamics of continua. Springer, New York

Jackson JD (1999) Classical electrodynamics. Wiley, New York

Kofod G, Sommer-Larsen P, Kornbluh R, Perline R (2003) Actuation response of polyacrylate dielectric elastomer. J Intell Mater Syst Struct 14:787–793

McMeeking RM, Landis CM (2005) Electrostatic forces and stored energy for deformable dielectric materials. J Appl Mech 72:581–590

Ogden RW, Steigmann DJ (eds) (2011) Mechanics and electrodynamics of magneto- and electro-elastic materials. Springer, Vienna

Pao YH (1978) Electromagnetic forces in deformable continua. In: Nemat-Nasser S (ed) Mechanics today. Pitman, Bath

Suo Z (2010) Theory of dielectric elastomers. Acta Mech Solid Sinica 23:549–578

Toupin RA (1956) The elastic dielectric. J Ration Mech Anal 5:849–914

Volokh KY (2012) On electromechanical coupling in elastomers. J Appl Mech 79:044507

Wissler M, Mazza E (2007) Mechanical behavior of an acrylic elastomer used in dielectric elastomer actuators. Sens Actuators A 134:494–504

Chapter 10
Viscoelasticity

Rubberlike materials and soft biological tissues can exhibit a time-dependent response. For example, stresses can decrease under the constant strains—*stress relaxation*—or strains can increase under the constant stresses—*creep*. Such phenomena are usually related to *viscosity*, which is a fluid-like property of materials. As usual in the case of soft materials, geometrical and physical nonlinearities make the well-known theory of linear viscoelasticity essentially useless. Strongly nonlinear theory of viscoelasticity is required and presented in this chapter.

10.1 Coupling of Elasticity and Fluidity

Deformation of materials can be provisionally divided in *elastic* and *inelastic*. The latter deformation can also be called *flow* to emphasize the fluid-like behavior of material.[1] Actually, all materials, with no exception, posses both modes of deformation. It is a matter of time under observation to define material as solid or liquid. This notion has biblical roots—Fig. 10.1.

For those readers who are too busy to track the flow of mountains it might be practical to observe mechanical behavior of *silly putty* (silicone polymer) that can jump as a rubber ball on the time-scale of seconds and flow like liquid on the time-scale of minutes.

It is instructive to start developing a general theoretical framework for elasticity-fluidity with a *rheological model*. The purpose of a rheological model is to create a primitive prototype of a three-dimensional theory. Though different tensorial formulations could be proposed for the same toy prototype they would share similar

---

[1] *Inelastic* deformation are not necessarily related to flow per se—they can be deformations resulted from growth of living tissues, for example.

© Springer Science+Business Media Singapore 2016
K. Volokh, *Mechanics of Soft Materials*, DOI 10.1007/978-981-10-1599-1_10

Fig. 10.1 "The mountains
flowed before the Lord" (The
Book of Judges)

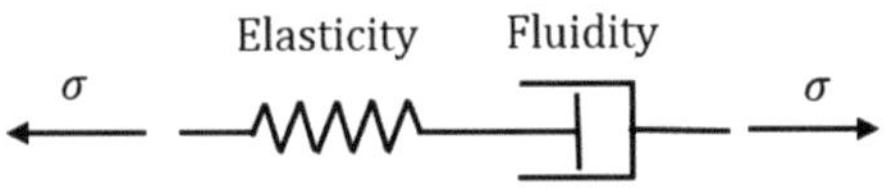

"הָרִים נָזְלוּ מִפְּנֵי יְהֹוָה"

Fig. 10.2 Rheological
(Maxwell) model of
elasticity and fluidity

Elasticity   Fluidity

qualitative features. Specifically, we choose the successively joined spring and dash-pot elements shown in Fig. 10.2 as a rheological model of elasticity–fluidity.

Here the spring element is related to elasticity while the dashpot element is related to fluidity. It is crucial for a three-dimensional tensorial formulation, which can stem from the rheological model, that stresses in elasticity and fluidity are equal because the spring and dashpot elements are joined *in series*.

10.1.1 Kinematics

We remind the reader that various approaches exist to describe deformations of the elements of the rheological model in the case of three-dimensional theory. We choose the following additive elastic–inelastic decomposition of the velocity gradient, which is arguably the most appealing physically,

$$\mathbf{L} = \mathrm{grad}\,\mathbf{v} = \mathbf{L_S} + \mathbf{L_F}, \tag{10.1}$$

where we use subscripts "S" for elastic Solid-like behavior and "F" for inelastic Fluid-like behavior.

The choice of the velocity gradient for a description of kinematics is natural in the case of flow.

We further decompose the elastic and inelastic parts of the velocity gradient into symmetric and skew tensors as follows

$$\mathbf{L_S} = \mathbf{D_S} + \mathbf{W_S}, \quad \mathbf{D_S} = \frac{1}{2}(\mathbf{L_S} + \mathbf{L_S^T}), \quad \mathbf{W_S} = \frac{1}{2}(\mathbf{L_S} - \mathbf{L_S^T}), \tag{10.2}$$

and

$$\mathbf{L_F} = \mathbf{D_F} + \mathbf{W_F}, \quad \mathbf{D_F} = \frac{1}{2}(\mathbf{L_F} + \mathbf{L_F^T}), \quad \mathbf{W_F} = \frac{1}{2}(\mathbf{L_F} - \mathbf{L_F^T}). \tag{10.3}$$

Here $\mathbf{D_S}$ and $\mathbf{D_F}$ are the elastic and inelastic deformation rate tensors accordingly; and $\mathbf{W_S}$ and $\mathbf{W_F}$ are the elastic and inelastic spin tensors accordingly.

We assume that the inelastic spin is zero

$$\mathbf{W_F} = \mathbf{0}, \tag{10.4}$$

and, consequently,

$$\mathbf{L} = \mathbf{L}_S + \mathbf{D}_F. \tag{10.5}$$

Decomposition (10.5) is a mathematical expression of the separate kinematic response of the spring and dashpot elements of the rheological model. We emphasize that assumption (10.4) corresponds to the isotropic material response. If the response is anisotropic then a constitutive equation for the inelastic spin should be defined.

10.1.2 Elasticity

The constitutive law describing the elastic behavior of the spring in the rheological model is a generalization of the three-dimensional isotropic hyperelastic solid discussed in Chap. 4

$$\boldsymbol{\sigma} = 2I_3^{-1/2}(I_3\psi_3\mathbf{1} + (\psi_1 + I_1\psi_2)\mathbf{B}_S - \psi_2\mathbf{B}_S^2), \tag{10.6}$$

where ψ is the elastic strain energy function.

Invariants are

$$I_1 = \text{tr}\mathbf{B}_S, \quad 2I_2 = (\text{tr}\mathbf{B}_S)^2 - \text{tr}(\mathbf{B}_S^2), \quad I_3 = \det \mathbf{B}_S, \tag{10.7}$$

and

$$\psi_j \equiv \frac{\partial\psi(I_1, I_2, I_3)}{\partial I_j}.$$

Symmetric elastic strain $\mathbf{B}_S = \mathbf{B}_S^T$ is not, generally, the left Cauchy–Green tensor used in Chap. 4 and we postpone its definition to Sect. 10.1.4.

10.1.3 Fluidity (Inelasticity)

The constitutive law describing inelastic behavior of the dashpot element in the rheological model—the *flow rule*—is a generalization of the *Reiner–Rivlin model* for isotropic fluid

$$\boldsymbol{\sigma} = \beta_1\mathbf{1} + \beta_2\mathbf{D}_F + \beta_3\mathbf{D}_F^2, \tag{10.8}$$

where the response *functionals* can depend on the history of the deformation

$$\beta_j(\boldsymbol{\sigma}, \mathbf{D}, \mathbf{D}_F).$$

We emphasize that constitutive equation (10.8) is *history-dependent* and *implicit*.

Equation (10.8) presents a description of flow while it is argued that an additional *yield condition* should be obeyed during the inelastic deformation

$$\gamma(\boldsymbol{\sigma}, \mathbf{D_F}) = 0. \tag{10.9}$$

The physical basis for the yield condition (10.9) is open for discussion. Amazingly, the yield condition (similarly to the condition of incompressibility) can be a blessing for an analytical solution or it can be a pain in the neck for a numerical procedure. The rate of the yield constraint (10.9) is often used to derive the so-called inelastic multiplier accounting for the history of inelastic deformations.

10.1.4   Evolution Equation

Previously, we used tensors $\mathbf{B_S}$ and $\mathbf{D_F}$ as kinematic variables describing elasticity and fluidity respectively. The purpose of this subsection is to establish a connection between them. The idea of this connection comes from the evolution equation that the left Cauchy–Green tensor $\mathbf{B} = \mathbf{FF}^T$ should obey

$$
\begin{aligned}
\dot{\mathbf{B}} &= \dot{\mathbf{F}}\mathbf{F}^T + \mathbf{F}\dot{\mathbf{F}}^T \\
&= \dot{\mathbf{F}}(\mathbf{F}^{-1}\mathbf{F})\mathbf{F}^T + \mathbf{F}(\mathbf{F}^{-1}\mathbf{F})^T\dot{\mathbf{F}}^T \\
&= (\dot{\mathbf{F}}\mathbf{F}^{-1})(\mathbf{FF}^T) + (\mathbf{FF}^T)(\mathbf{F}^{-T}\dot{\mathbf{F}}^T) \\
&= \mathbf{LB} + \mathbf{BL}^T,
\end{aligned}
\tag{10.10}
$$

where (2.49) is used.

Thus, the evolving left Cauchy–Green tensor is a solution of the *initial value problem* (IVP)

$$
\begin{aligned}
\dot{\mathbf{B}} - \mathbf{LB} - \mathbf{BL}^T &= \mathbf{0}, \\
\mathbf{B}(t = 0) &= \mathbf{1}.
\end{aligned}
\tag{10.11}
$$

By analogy with $\mathbf{B}$, we assume now that the elastic strain $\mathbf{B_S}$ is *defined* as a solution of a similar IVP

$$
\begin{aligned}
\dot{\mathbf{B}}_S - \mathbf{L_S}\mathbf{B_S} - \mathbf{B_S}\mathbf{L}_S^T &= \mathbf{0}, \\
\mathbf{B_S}(t = 0) &= \mathbf{1}.
\end{aligned}
\tag{10.12}
$$

In the absence of inelastic fluid-like deformations both IVPs coincide and $\mathbf{B_S} = \mathbf{B} = \mathbf{FF}^T$.

Substitution of (10.5) in (10.12)$_1$ yields evolution equation in the form

$$\dot{\mathbf{B}}_S - \mathbf{LB_S} - \mathbf{B_S}\mathbf{L}^T + \mathbf{D_F}\mathbf{B_S} + \mathbf{B_S}\mathbf{D_F} = \mathbf{0}. \tag{10.13}$$

We note that the first three terms on the left hand side of the equation present the Oldroyd objective rate $(2.66)_3$ of $\mathbf{B}_S$

$$\dot{\mathbf{B}}_S - \mathbf{L}\mathbf{B}_S - \mathbf{B}_S\mathbf{L}^T = \overset{\diamond}{\mathbf{B}}_S. \tag{10.14}$$

Besides, tensors $\mathbf{B}_S$ and $\mathbf{D}_F$ are coaxial by virtue of constitutive laws (10.6) and (10.8) and, consequently,

$$\mathbf{D}_F\mathbf{B}_S = \mathbf{B}_S\mathbf{D}_F. \tag{10.15}$$

Thus, the evolution equation can be written in the compact form

$$\overset{\diamond}{\mathbf{B}}_S = -2\mathbf{D}_F\mathbf{B}_S. \tag{10.16}$$

The evolution equation provides the connection between the elastic $\mathbf{B}_S$ and inelastic $\mathbf{D}_F$ deformation measures.

We note, finally, that the evolution equation for elastic deformations is invariant under the superposed rigid body motion. Indeed, let us designate $\mathbf{Q}$ a proper orthogonal tensor describing the superimposed rigid body rotation. Then designating the rotated quantities with asterisk we have

$$\begin{aligned}
\mathbf{B}_S^* &= \mathbf{Q}\mathbf{B}_S\mathbf{Q}^T, \\
\mathbf{L}^* &= \mathbf{Q}\mathbf{L}\mathbf{Q}^T + \dot{\mathbf{Q}}\mathbf{Q}^T = \mathbf{Q}(\mathbf{L}_S + \mathbf{D}_F)\mathbf{Q}^T + \dot{\mathbf{Q}}\mathbf{Q}^T = \mathbf{L}_S^* + \mathbf{D}_F^*,
\end{aligned} \tag{10.17}$$

where

$$\begin{aligned}
\mathbf{L}_S^* &= \mathbf{Q}\mathbf{L}_S\mathbf{Q}^T + \dot{\mathbf{Q}}\mathbf{Q}^T, \\
\mathbf{D}_F^* &= \mathbf{Q}\mathbf{D}_F\mathbf{Q}^T.
\end{aligned} \tag{10.18}$$

With account of these transformations we get by a direct calculation

$$\dot{\mathbf{B}}_S^* - \mathbf{L}_S^*\mathbf{B}_S^* - \mathbf{B}_S^*\mathbf{L}_S^{*T} = \mathbf{Q}(\dot{\mathbf{B}}_S - \mathbf{L}_S\mathbf{B}_S - \mathbf{B}_S\mathbf{L}_S^T)\mathbf{Q}^T. \tag{10.19}$$

10.1.5 Thermodynamic Restrictions

Let us neglect thermal effects and examine the dissipation inequality in the form

$$D_{\text{int}} = \boldsymbol{\sigma} : \mathbf{D} - I_3^{-1/2}\dot{\psi} \geq 0. \tag{10.20}$$

Decomposing the deformation rate tensor additively in elastic and flow parts in accordance with (10.5) and calculating the free energy increment we get

$$D_{\text{int}} = \boldsymbol{\sigma} : \mathbf{D}_S + \boldsymbol{\sigma} : \mathbf{D}_F - I_3^{-1/2}\frac{\partial\psi}{\partial\mathbf{B}_S} : \dot{\mathbf{B}}_S \geq 0. \tag{10.21}$$

The third term on the left-hand side of this inequality can be further calculated
with the help of the evolution equation (10.12) as follows

$$
\frac{\partial \psi}{\partial \mathbf{B}_S} : \dot{\mathbf{B}}_S = \frac{\partial \psi}{\partial \mathbf{B}_S} : \mathbf{L}_S \mathbf{B}_S + \frac{\partial \psi}{\partial \mathbf{B}_S} : \mathbf{B}_S \mathbf{L}_S^T = 2 \frac{\partial \psi}{\partial \mathbf{B}_S} \mathbf{B}_S : \mathbf{L}_S = 2 \frac{\partial \psi}{\partial \mathbf{B}_S} \mathbf{B}_S : \mathbf{D}_S.
$$

$$(10.22)$$

Substitution of (10.22) in (10.21) yields

$$
D_{\text{int}} = (\sigma - 2 I_3^{-1/2} \frac{\partial \psi}{\partial \mathbf{B}_S} \mathbf{B}_S) : \mathbf{D}_S + \sigma : \mathbf{D}_F \geq 0. \tag{10.23}
$$

We note that the expression in the parentheses equals zero by virtue of the hyperelastic constitutive law

$$
\sigma = 2 I_3^{-1/2} \frac{\partial \psi}{\partial \mathbf{B}_S} \mathbf{B}_S = 2 I_3^{-1/2} (I_3 \psi_3 \mathbf{1} + (\psi_1 + I_1 \psi_2) \mathbf{B}_S - \psi_2 \mathbf{B}_S^2). \tag{10.24}
$$

Thus, the dissipation inequality reduces to

$$
D_{\text{int}} = \sigma : \mathbf{D}_F \geq 0. \tag{10.25}
$$

Substituting flow rule (10.8) for the Cauchy stress we obtain, alternatively,

$$
D_{\text{int}} = \beta_1 \text{tr}(\mathbf{D}_F) + \beta_2 \text{tr}(\mathbf{D}_F^2) + \beta_3 \text{tr}(\mathbf{D}_F^3) \geq 0. \tag{10.26}
$$

In this way, the dissipation inequality imposes a restriction on the inelastic
response functionals β_j and the processes of flow.

10.2 Viscoelastic Framework for "Standard Solid"

In this section we develop a nonlinear theory of viscoelasticity based on the rheological model of the "standard solid", in which nonlinear spring A is parallel to
nonlinear spring and dashpot B joined in series—Fig. 10.3.

Fig. 10.3 Rheological
model of the "standard solid"

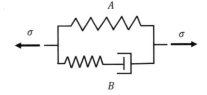

10.2.1   Constitutive Equations

In this case, the constitutive law for the Cauchy stress takes form

$$\boldsymbol{\sigma} = \boldsymbol{\sigma}_A + \boldsymbol{\sigma}_B, \tag{10.27}$$

where

$$\boldsymbol{\sigma}_A = 2I_{A3}^{-1/2}\frac{\partial \psi_A}{\partial \mathbf{B}}\mathbf{B} = 2I_{A3}^{-1/2}(I_{A3}\psi_{A3}\mathbf{1} + (\psi_{A1} + I_{A1}\psi_{A2})\mathbf{B} - \psi_{A2}\mathbf{B}^2), \tag{10.28}$$

and

$$\boldsymbol{\sigma}_B = 2I_{B3}^{-1/2}\frac{\partial \psi_B}{\partial \mathbf{B}_B}\mathbf{B}_B = 2I_{B3}^{-1/2}(I_{B3}\psi_{B3}\mathbf{1} + (\psi_{B1} + I_{B1}\psi_{B2})\mathbf{B}_B - \psi_2\mathbf{B}_B^2), \tag{10.29}$$

where $\mathbf{B}_A = \mathbf{F}\mathbf{F}^{\mathrm{T}} = \mathbf{B}$ is the left Cauchy–Green tensor; $\mathbf{B}_B = \mathbf{B}_B^{\mathrm{T}}$; ψ_A and ψ_B are strain energies of the springs; $\psi_{As} = \partial\psi_A/\partial I_{As}$ and $\psi_{Bs} = \partial\psi_B/\partial I_{Bs}$; and the principal invariants are

$$I_{A1} = \mathrm{tr}\mathbf{B}, \quad 2I_{A2} = (\mathrm{tr}\mathbf{B})^2 - \mathrm{tr}(\mathbf{B}^2), \quad I_{A3} = \det\mathbf{B}, \tag{10.30}$$

and

$$I_{B1} = \mathrm{tr}\mathbf{B}_B, \quad 2I_{B2} = (\mathrm{tr}\mathbf{B}_B)^2 - \mathrm{tr}(\mathbf{B}_B^2), \quad I_{B3} = \det\mathbf{B}_B. \tag{10.31}$$

The flow rule for dashpot B can be written in the following general form

$$\boldsymbol{\sigma}_B = \beta_1\mathbf{1} + \beta_2\mathbf{D}_B + \beta_3\mathbf{D}_B^2, \tag{10.32}$$

where β_j is a function(al), generally, depending on stresses and strains.

10.2.2   Kinematics

We note that A and B parts of the rheological model have different stresses $\boldsymbol{\sigma}_A$ and $\boldsymbol{\sigma}_B$ but velocity gradients are equal because of the parallel architecture of the "standard solid"

$$\mathbf{L}_A = \mathbf{L}_B = \mathbf{L} = \mathrm{grad}\mathbf{v}. \tag{10.33}$$

The B part of the model has the successive connection of the spring and dashpot and, consequently, we can use the additive decomposition as in (10.5)

$$\mathbf{L}_B = \mathbf{L}_S + \mathbf{D}_F, \tag{10.34}$$

where

$$\mathbf{D}_F = \mathbf{D}_B. \tag{10.35}$$

Then, the relationship between the strain measure $\mathbf{B}_B$ in spring B and the strain rate measure $\mathbf{D}_B$ in dashpot B can be defined analogously to (10.16) as follows

$$\overset{\diamond}{\mathbf{B}}_B = -2\mathbf{D}_B\mathbf{B}_B,$$
$$\mathbf{B}_B(t=0) = \mathbf{1}, \tag{10.36}$$

where

$$\overset{\diamond}{\mathbf{B}}_B = \dot{\mathbf{B}}_B - \mathbf{L}\mathbf{B}_B - \mathbf{B}_B\mathbf{L}^{\mathrm{T}} \tag{10.37}$$

is the Oldroyd objective rate of the B spring strain measure.

In the case of $\mathbf{D}_B = \mathbf{0}$ we have two parallel springs and $\mathbf{B}_B = \mathbf{B}_A = \mathbf{B}$.

10.2.3   Thermodynamic Restrictions

The second law of thermodynamics imposes restrictions on the constitutive laws. We write the dissipation inequality in the following form

$$D_{\text{int}} = (\boldsymbol{\sigma}_A + \boldsymbol{\sigma}_B) : \mathbf{D} - I_{A3}^{-1/2}\frac{\partial\psi_A}{\partial\mathbf{B}} : \dot{\mathbf{B}} - I_{B3}^{-1/2}\frac{\partial\psi_B}{\partial\mathbf{B}_B} : \dot{\mathbf{B}}_B \geq 0. \tag{10.38}$$

The strain increments can be calculated as follows

$$\dot{\mathbf{B}} = \mathbf{L}\mathbf{B} + \mathbf{B}\mathbf{L}^{\mathrm{T}}, \tag{10.39}$$

and

$$\dot{\mathbf{B}}_B = (\mathbf{L} - \mathbf{D}_B)\mathbf{B}_B + \mathbf{B}_B(\mathbf{L} - \mathbf{D}_B)^{\mathrm{T}}. \tag{10.40}$$

Consequently, we have

$$\frac{\partial\psi_A}{\partial\mathbf{B}} : \dot{\mathbf{B}} = \frac{\partial\psi_A}{\partial\mathbf{B}} : (\mathbf{L}\mathbf{B} + \mathbf{B}\mathbf{L}^{\mathrm{T}})$$
$$= 2\frac{\partial\psi_A}{\partial\mathbf{B}}\mathbf{B} : \mathbf{L}$$
$$= 2\frac{\partial\psi_A}{\partial\mathbf{B}}\mathbf{B} : \mathbf{D}, \tag{10.41}$$

and

$$\frac{\partial \psi_B}{\partial \mathbf{B}_B} : \dot{\mathbf{B}}_B = \frac{\partial \psi_B}{\partial \mathbf{B}_B} : [(\mathbf{L} - \mathbf{D}_B)\mathbf{B}_B + \mathbf{B}_B(\mathbf{L} - \mathbf{D}_B)^{\mathrm{T}}]$$

$$= 2\frac{\partial \psi_B}{\partial \mathbf{B}_B}\mathbf{B}_B : (\mathbf{L} - \mathbf{D}_B)$$

$$= 2\frac{\partial \psi_B}{\partial \mathbf{B}_B}\mathbf{B}_B : \mathbf{D} - 2\frac{\partial \psi_B}{\partial \mathbf{B}_B}\mathbf{B}_B : \mathbf{D}_B. \tag{10.42}$$

Substitution of (10.41) and (10.42) in (10.38) yields

$$\left(\boldsymbol{\sigma}_A - 2I_{A3}^{-1/2}\frac{\partial \psi_A}{\partial \mathbf{B}}\mathbf{B} + \boldsymbol{\sigma}_B - 2I_{B3}^{-1/2}\frac{\partial \psi_B}{\partial \mathbf{B}_B}\mathbf{B}_B\right) : \mathbf{D} + 2I_{B3}^{-1/2}\frac{\partial \psi_B}{\partial \mathbf{B}_B}\mathbf{B}_B : \mathbf{D}_B \geq 0. \tag{10.43}$$

By virtue of the hyperelastic constitutive equations (10.28) and (10.29) the dissipation inequality further reduces to

$$D_{\mathrm{int}} = \boldsymbol{\sigma}_B : \mathbf{D}_B \geq 0. \tag{10.44}$$

Substituting the flow rule (10.32) in the latter inequality we get the final thermodynamic restriction

$$D_{\mathrm{int}} = \beta_1 \mathrm{tr}\mathbf{D}_B + \beta_2 \mathrm{tr}\mathbf{D}_B^2 + \beta_3 \mathrm{tr}\mathbf{D}_B^3 \geq 0. \tag{10.45}$$

10.3 Incompressibility

We enforce incompressibility in the constitutive laws for the springs (10.28) and (10.29) as follows

$$\boldsymbol{\sigma}_A = -\Pi_A \mathbf{1} + 2(\psi_{A1} + I_{A1}\psi_{A2})\mathbf{B} - 2\psi_{A2}\mathbf{B}^2, \quad \det\mathbf{B} = 1, \tag{10.46}$$

and

$$\boldsymbol{\sigma}_B = -\Pi_B \mathbf{1} + 2(\psi_{B1} + I_{B1}\psi_{B2})\mathbf{B}_B - 2\psi_2\mathbf{B}_B^2, \quad \det\mathbf{B}_B = 1, \tag{10.47}$$

where Π_A and Π_B are the undefined Lagrange parameters.

Let us introduce incompressibility in the constitutive laws for the dashpot. Consider, first, the general kinematic condition of incompressibility $J = 1$. Differentiating it with respect to time and accounting for (1.56) and (2.49) we get

$$\dot{J} = \frac{\partial J}{\partial \mathbf{F}} : \dot{\mathbf{F}} = J\mathbf{F}^{-\mathrm{T}} : \dot{\mathbf{F}} = J\mathrm{tr}\mathbf{L} = J\mathrm{tr}\mathbf{D} = 0. \tag{10.48}$$

By analogy with this general kinematic restriction we require for the dashpot deformation rate

$$\text{tr}\mathbf{D}_B = \mathbf{1} : \mathbf{D}_B = 0. \tag{10.49}$$

This equation can be interpreted as a stress power where the workless spherical stress can be multiplied by an arbitrary multiplier: $\Xi\mathbf{1}$; and the flow rule (10.32) takes form

$$\boldsymbol{\sigma}_B = (\beta_1 + \Xi)\mathbf{1} + \beta_2\mathbf{D}_B + \beta_3\mathbf{D}_B^2. \tag{10.50}$$

For a wide class of materials it is possible to set

$$\beta_3 = 0, \tag{10.51}$$

and, consequently, we get

$$\boldsymbol{\sigma}_B = (\beta_1 + \Xi)\mathbf{1} + \beta_2\mathbf{D}_B. \tag{10.52}$$

Taking trace on both sides of this equation we find

$$\beta_1 + \Xi = \frac{1}{3}\text{tr}\boldsymbol{\sigma}_B, \tag{10.53}$$

and, therefore, we have finally

$$\boldsymbol{\sigma}_B = \frac{1}{3}(\text{tr}\boldsymbol{\sigma}_B)\mathbf{1} + \beta_2\mathbf{D}_B. \tag{10.54}$$

This special flow rule can be interpreted as a particular case of (10.32) for the following choice of parameters

$$\beta_1 = \frac{1}{3}\text{tr}\boldsymbol{\sigma}_B, \quad \beta_3 = 0. \tag{10.55}$$

This flow rule is often written in the reversed form

$$\mathbf{D}_B = \frac{1}{\beta_2}\text{dev}\boldsymbol{\sigma}_B, \tag{10.56}$$

where the *deviatoric* stress tensor is introduced by using the following definition

$$\text{dev}\boldsymbol{\sigma}_B = \boldsymbol{\sigma}_B - \frac{1}{3}(\text{tr}\boldsymbol{\sigma}_B)\mathbf{1}. \tag{10.57}$$

Substitution of (10.54) in (10.44) yields

$$D_{\text{int}} = \beta_2\text{tr}\mathbf{D}_B^2 \geq 0. \tag{10.58}$$

This dissipation inequality can be obeyed by imposing the following restrictions on the viscosity parameter

$$\beta_2 > 0. \tag{10.59}$$

10.4 Uniaxial Tension

Consider uniaxial uniform tension in direction $\mathbf{e}_1$. In this case we have the following kinematic variables

$$\begin{aligned}
\mathbf{y} &= \lambda x_1 \mathbf{e}_1 + \lambda^{-1/2}(x_2 \mathbf{e}_2 + x_3 \mathbf{e}_3), \\
\mathbf{F} &= \lambda \mathbf{e}_1 \otimes \mathbf{e}_1 + \lambda^{-1/2}(\mathbf{e}_2 \otimes \mathbf{e}_2 + \mathbf{e}_3 \otimes \mathbf{e}_3), \\
\mathbf{B} &= \lambda^2 \mathbf{e}_1 \otimes \mathbf{e}_1 + \lambda^{-1}(\mathbf{e}_2 \otimes \mathbf{e}_2 + \mathbf{e}_3 \otimes \mathbf{e}_3), \\
\mathbf{B}_B &= \lambda_B^2 \mathbf{e}_1 \otimes \mathbf{e}_1 + \lambda_B^{-1}(\mathbf{e}_2 \otimes \mathbf{e}_2 + \mathbf{e}_3 \otimes \mathbf{e}_3), \\
\mathbf{D}_B &= d_B \mathbf{e}_1 \otimes \mathbf{e}_1 - \frac{d_B}{2}(\mathbf{e}_2 \otimes \mathbf{e}_2 + \mathbf{e}_3 \otimes \mathbf{e}_3),
\end{aligned} \tag{10.60}$$

which obey the incompressibility conditions $\det\mathbf{B} = 1$, $\det\mathbf{B}_B = 1$, $\mathrm{tr}\mathbf{D}_B = 0$.
    The velocity and its gradient are

$$\begin{aligned}
\mathbf{v} &= \dot{\lambda} x_1 \mathbf{e}_1 - \frac{\dot{\lambda}}{2\lambda^{3/2}}(x_2 \mathbf{e}_2 + x_3 \mathbf{e}_3) = \frac{\dot{\lambda}}{2\lambda}(2y_1 \mathbf{e}_1 - y_2 \mathbf{e}_2 - y_3 \mathbf{e}_3), \\
\mathbf{L} &= \mathrm{grad}\mathbf{v} = \frac{\dot{\lambda}}{2\lambda}(2\mathbf{e}_1 \otimes \mathbf{e}_1 - \mathbf{e}_2 \otimes \mathbf{e}_2 - \mathbf{e}_3 \otimes \mathbf{e}_3) = \mathbf{D}.
\end{aligned} \tag{10.61}$$

We note that strains are homogeneous: $\mathrm{grad}\mathbf{B}_B = \mathbf{0}$; and, the full material time derivative equals the partial time derivative: $\dot{\mathbf{B}}_B = \partial\mathbf{B}_B/\partial t$. Therefore, we get the time derivative of $\mathbf{B}_B$ in the form

$$\dot{\mathbf{B}}_B = 2\lambda_B \dot{\lambda}_B \mathbf{e}_1 \otimes \mathbf{e}_1 - \lambda_B^{-2} \dot{\lambda}_B(\mathbf{e}_2 \otimes \mathbf{e}_2 + \mathbf{e}_3 \otimes \mathbf{e}_3). \tag{10.62}$$

In view of these kinematic simplifications, (10.36) reduces to one scalar kinematic evolution equation

$$\dot{\lambda}_B - \dot{\lambda}\lambda^{-1}\lambda_B + d_B \lambda_B = 0, \quad \lambda_B(t = 0) = 1. \tag{10.63}$$

Constitutive equations for springs take forms

$$\boldsymbol{\sigma}_A = \sigma_{A1} \mathbf{e}_1 \otimes \mathbf{e}_1 + \sigma_{A2}(\mathbf{e}_2 \otimes \mathbf{e}_2 + \mathbf{e}_3 \otimes \mathbf{e}_3), \tag{10.64}$$

where
$$\sigma_{A1} = -\Pi_A + 2(\psi_{A1} + 2\lambda^{-1}\psi_{A2})\lambda^2,$$
$$\sigma_{A2} = -\Pi_A + 2(\psi_{A1} + (\lambda^2 + \lambda^{-1})\psi_{A2})\lambda^{-1}, \tag{10.65}$$

and
$$\boldsymbol{\sigma}_B = \sigma_{B1}\mathbf{e}_1 \otimes \mathbf{e}_1 + \sigma_{B2}(\mathbf{e}_2 \otimes \mathbf{e}_2 + \mathbf{e}_3 \otimes \mathbf{e}_3), \tag{10.66}$$

where
$$\sigma_{B1} = -\Pi_B + 2(\psi_{B1} + 2\lambda_B^{-1}\psi_{B2})\lambda_B^2,$$
$$\sigma_{B2} = -\Pi_B + 2(\psi_{B1} + (\lambda_B^2 + \lambda_B^{-1})\psi_{B2})\lambda_B^{-1}. \tag{10.67}$$

Normal stresses in directions $\mathbf{e}_2$ and $\mathbf{e}_3$ are zero and we find the Lagrange parameters
$$\Pi_A = 2(\psi_{A1} + (\lambda^2 + \lambda^{-1})\psi_{A2})\lambda^{-1},$$
$$\Pi_B = 2(\psi_{B1} + (\lambda_B^2 + \lambda_B^{-1})\psi_{B2})\lambda_B^{-1}. \tag{10.68}$$

Substituting the Lagrange parameters back in the stress formulas we get finally
$$\sigma_{A1} = 2(\lambda^2 - \lambda^{-1})(\psi_{A1} + \lambda^{-1}\psi_{A2}),$$
$$\sigma_{B1} = 2(\lambda_B^2 - \lambda_B^{-1})(\psi_{B1} + \lambda_B^{-1}\psi_{B2}).$$

The flow rule reduces to
$$d_B = \frac{2}{3\beta_2}\sigma_{B1} = \frac{4}{3\beta_2}(\lambda_B^2 - \lambda_B^{-1})(\psi_{B1} + \lambda_B^{-1}\psi_{B2}). \tag{10.69}$$

Substitution of flow rule (10.69) in the evolution equation (10.63) yields
$$\dot{\lambda}_B = \dot{\lambda}\lambda^{-1}\lambda_B - \frac{4\lambda_B}{3\beta_2}(\lambda_B^2 - \lambda_B^{-1})(\psi_{B1} + \lambda_B^{-1}\psi_{B2}) = 0, \quad \lambda_B(t = 0) = 1. \tag{10.70}$$

The stress–stretch curve, $\sigma \sim \lambda$, is given by the following equation
$$\sigma = \sigma_{A1} + \sigma_{B1} = 2(\lambda^2 - \lambda^{-1})(\psi_1 + \lambda^{-1}\psi_2) + 2(\lambda_B^2 - \lambda_B^{-1})(\psi_{B1} + \lambda_B^{-1}\psi_{B2}), \tag{10.71}$$

in which λ_B is calculated from (10.70) for given λ and $\dot{\lambda}$.

In the case of steady stretching we have
$$\lambda = 1 + \dot{\lambda}t, \quad \dot{\lambda} = \text{constant}, \tag{10.72}$$

and we can replace the time derivative as follows
$$\frac{d}{dt} = \dot{\lambda}\frac{d}{d\lambda}, \tag{10.73}$$

Fig. 10.4 Cauchy stress [MPa] versus stretch in uniaxial tension. Comparison of the theory (*solid lines*) and experiment (*triangles*) under strain rates increasing from the bottom to the top: 0.001 1/s; 150 1/s; 450 1/s

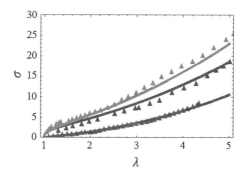

With account of this replacement, (10.70) takes form

$$\frac{d\lambda_B}{d\lambda} = \lambda^{-1}\lambda_B - \frac{4\lambda_B}{3\eta_2\lambda}(\lambda_B^2 - \lambda_B^{-1})(\psi_{B1} + \lambda_B^{-1}\psi_{B2}) = 0, \quad \lambda_B(\lambda = 1) = 1. \tag{10.74}$$

We use the experimental data reported by Hoo Fatt and Ouyang (2008) for uniaxial tension test—Fig. 10.4—at various strain rates to further specialize and calibrate the model as follows

$$\psi_A = \frac{\mu_A}{\gamma_A}(I_{A1} - 3)^{\gamma_A}, \quad \psi_B = \frac{\mu_B}{\gamma_B}(I_{B1} - 3)^{\gamma_B}, \tag{10.75}$$

$$\mu_A = 0.195 \text{ MPa}, \quad \mu_B = 1.25 \text{ MPa}, \quad \gamma_A = 1.02, \quad \gamma_B = 0.67246.$$

Unfortunately, the constant viscosity parameter β_2 is not enough to fit the experimental data and we define it as a function[2]

$$\beta_2(I_1, I_{B1}) = \beta_2^*(I_1)\beta_2^{**}(I_{B1}), \tag{10.76}$$

where

$$\beta_2^*(I_1) = c_1(1 - \exp[c_2(I_1 - 3)]) + c_3,$$
$$\beta_2^{**}(I_{B1}) = c_4 I_{B1}^3 + c_5 I_{B1}^2 + c_6 I_{B1} + c_7, \tag{10.77}$$

and

$$c_1 = 0.5 \text{ MPa} \cdot \text{s}, \quad c_2 = -0.13726, \quad c_3 = 0.0939 \text{ MPa} \cdot \text{s},$$
$$c_4 = -2.54 \cdot 10^{-6}, \quad c_5 = 0.02, \quad c_6 = -0.00146, \quad c_7 = 0.1.$$

Viscosity functions $\beta_2^*(I_1)$ and $\beta_2^{**}(I_{B1})$ are presented graphically in Fig. 10.5.

---

[2]We follow Hoo Fatt and Ouyang (2008) yet with slightly different constants.

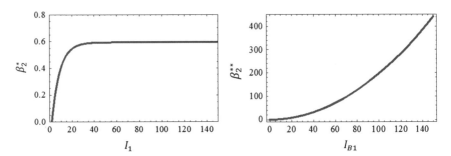

Fig. 10.5 Viscosity functions β_2^* [MPa · s] and β_2^{**}

10.5   Integration of Constitutive Equations

The constitutive theory developed in the previous sections can be implemented in the existing *finite element* (FE) codes by using the user-defined subroutines. The latter subroutines, however, require definitions of numerical schemes for the time integration of the constitutive equations. Normally, a FE program provides all variables at time t_n and the deformation gradient for time t_{n+1}. The user-defined subroutine should provide an algorithm to update all variables at time t_{n+1}. The reader is advised to consult Simo and Hughes (1998) or Belytschko et al. (2000) for the general background. In this section we provide such an algorithm for the explicit constitutive update.

The hyperelastic constitutive laws and the flow rule do not need integration but Eq. (10.36) relating the B spring strain rate $\overset{\diamond}{\mathbf{B}}_B$ and the dashpot strain rate $\mathbf{D}_B$ should be integrated. The idea of the integration comes from the notion that the Oldroyd objective rate can be written with respect to *arbitrary* reference configuration $\mathbf{z}$ in the following form

$$\overset{\diamond}{\mathbf{B}}_B = \dot{\mathbf{B}}_B - \mathbf{L}\mathbf{B}_B - \mathbf{B}_B\mathbf{L}^{\mathrm{T}} = \mathbf{K}\left\{\frac{\partial}{\partial t}(\mathbf{K}^{-1}\mathbf{B}_B\mathbf{K}^{-\mathrm{T}})\right\}\mathbf{K}^{\mathrm{T}}, \qquad (10.78)$$

where

$$\mathbf{K} = \frac{\partial \mathbf{y}}{\partial \mathbf{z}},$$

$$\mathbf{L} = \frac{\partial \mathbf{v}}{\partial \mathbf{z}}\frac{\partial \mathbf{z}}{\partial \mathbf{y}} = \dot{\mathbf{K}}\mathbf{K}^{-1},$$

$$\dot{\mathbf{K}} = \frac{\partial \mathbf{v}}{\partial \mathbf{z}}, \qquad (10.79)$$

$$\dot{\mathbf{K}}^{-1} = -\mathbf{K}^{-1}\dot{\mathbf{K}}\mathbf{K}^{-1}.$$

Substitution of (10.78) in (10.36) yields

$$\mathbf{K}\left\{\frac{\partial}{\partial t}(\mathbf{K}^{-1}\mathbf{B}_B\mathbf{K}^{-T})\right\}\mathbf{K}^{T} = -\mathbf{D}_B\mathbf{B}_B - \mathbf{B}_B\mathbf{D}_B, \tag{10.80}$$

with the initial condition $\mathbf{B}_B(t=0) = \mathbf{1}$.

Using the simplest approximation of the time derivative on interval $[t_n, t_{n+1}]$ and with account of coaxiality, $\mathbf{D}_B\mathbf{B}_B = \mathbf{B}_B\mathbf{D}_B$, we get

$$\mathbf{K}^{-1}(t_{n+1})\mathbf{B}_B(t_{n+1})\mathbf{K}^{-T}(t_{n+1}) - \mathbf{K}^{-1}(t_n)\mathbf{B}_B(t_n)\mathbf{K}^{-T}(t_n)$$
$$= -2(t_{n+1} - t_n)\mathbf{K}^{-1}(t_n)\mathbf{D}_B(t_n)\mathbf{B}_B(t_n)\mathbf{K}^{-T}(t_n) \tag{10.81}$$

and

$$\mathbf{z} = \mathbf{y}(t_n),$$
$$\mathbf{K}(t_{n+1}) = \mathbf{F}(t_{n+1})\mathbf{F}^{-1}(t_n), \tag{10.82}$$
$$\mathbf{K}(t_n) = \mathbf{1}.$$

Substitution of (10.82)$_2$ in (10.81) yields

$$\mathbf{B}_B(t_{n+1}) = \mathbf{K}(t_{n+1})\{\mathbf{1} - 2(t_{n+1} - t_n)\mathbf{D}_B(t_n)\}\mathbf{B}_B(t_n)\mathbf{K}^{T}(t_{n+1}). \tag{10.83}$$

We assume now that variables $\mathbf{F}(t_n), \mathbf{B}(t_n), \mathbf{D}(t_n), \mathbf{B}_B(t_n), \mathbf{D}_B(t_n), \sigma_A(t_n), \sigma_B(t_n)$ are known. Besides, the deformation gradient $\mathbf{F}(t_{n+1})$ is computed from the momentum balance. Then, we update variables at time t_{n+1} as follows

$$\begin{aligned}
\mathbf{K}(t_{n+1}) &= \mathbf{F}(t_{n+1})\mathbf{F}^{-1}(t_n), \\
\dot{\mathbf{K}}(t_{n+1}) &= \frac{1}{t_{n+1} - t_n}\{\mathbf{K}(t_{n+1}) - \mathbf{1}\}, \\
\mathbf{L}(t_{n+1}) &= \dot{\mathbf{K}}(t_{n+1})\mathbf{K}^{-1}(t_{n+1}), \\
\mathbf{D}(t_{n+1}) &= \frac{1}{2}\{\mathbf{L}(t_{n+1}) + \mathbf{L}^{T}(t_{n+1})\}, \\
\mathbf{B}(t_{n+1}) &= \mathbf{F}(t_{n+1})\mathbf{F}^{T}(t_{n+1}), \\
\mathbf{B}_B(t_{n+1}) &= \mathbf{K}(t_{n+1})\{\mathbf{1} - 2(t_{n+1} - t_n)\mathbf{D}_B(t_{n+1})\}\mathbf{B}_B(t_n)\mathbf{K}^{T}(t_{n+1}),
\end{aligned} \tag{10.84}$$

and

$$\sigma_A(t_{n+1}) = 2I_{A3}^{-1/2}(t_{n+1})\{I_{A3}(t_{n+1})\psi_{A3}(t_{n+1})\mathbf{1} - \psi_{A2}(t_{n+1})\mathbf{B}^2(t_{n+1})$$
$$+ [\psi_{A1}(t_{n+1}) + I_{A1}(t_{n+1})\psi_{A2}(t_{n+1})]\mathbf{B}(t_{n+1})\}, \tag{10.85}$$

$$\sigma_B(t_{n+1}) = 2I_{B3}^{-1/2}(t_{n+1})\{I_{B3}(t_{n+1})\psi_{B3}(t_{n+1})\mathbf{1} - \psi_{B2}(t_{n+1})\mathbf{B}_B^2(t_{n+1})$$
$$+ [\psi_{B1}(t_{n+1}) + I_{B1}(t_{n+1})\psi_{B2}(t_{n+1})]\mathbf{B}_B(t_{n+1})\}, \tag{10.86}$$

and, finally,

$$\mathbf{D}_B(t_{n+1}) = \frac{2}{9\beta_1(t_{n+1})}[\mathrm{tr}\boldsymbol{\sigma}_B(t_{n+1})]\mathbf{1} + \frac{1}{\beta_2(t_{n+1})}\mathrm{dev}\boldsymbol{\sigma}_B(t_{n+1}), \qquad (10.87)$$

where $\beta_3 = 0$ is assumed.

References

Belytschko T, Liu WK, Moran B (2000) Nonlinear finite elements for continua and structures. Wiley, New York

Eckart C (1948) The thermodynamics of irreversible processes. IV. The theory of elasticity and anelasticity. Phys Rev 73:373–382

Hoo Fatt MS, Ouyang X (2008) Three-dimensional constitutive equations for styrene butadiene rubber at high strain rates. Mech Mat 40:1–16

Leonov AL (1976) Nonequilibrium thermodynamics and rheology of viscoelastic polymer media. Rheologica Acta 15:85–98

Simo JC, Hughes TJR (1998) Computational inelasticity. Springer, New York

Tanner RI, Walters K (1998) Rheology: an historical perspective. Elsevier, Amsterdam

Phan-Thien N (2013) Understanding viscoelasticity: an introduction to rheology. Springer, Berlin

Volokh KY (2013) An approach to elastoplasticity at large deformations. Euro J Mech A/Solids 39:153–162

Index

Printed in the United States
By Bookmasters